Implementing EDI

For a complete listing of the *Artech House Telecommuncations Library*,
turn to the back of this book

Implementing EDI

Mike Hendry

Artech House
Boston • London

Library of Congress Cataloging-in-Publication Data

Hendry, Mike
Implementing EDI/Mike Hendry
Includes bibliographical references and index
ISBN 0-89006-664-7
1. Computer networks 2. Electronic data interchange
TK5015.5.H465   1993
343.3–dc20

93-28374
CIP

© 1993 ARTECH HOUSE, INC.
685 Canton Street
Norwood, MA 02062

International Standard Book Number: 0-89006-664-7
Library of Congress Catalog Card Number: 93-28374

10 9 8 7 6 5 4 3 2 1

Contents

Part I

Chapter 1
Concept and Definitions

1.1 WHAT IS EDI?

EDI stands for "Electronic Data Interchange." The concept is very simple: the use of direct links between computers, even between computers on different sites, to transmit data that would otherwise usually be sent in printed form.

Although EDI is most often thought of as replacing standardized documents such as order forms, delivery notes, and invoices, the technique is highly flexible. EDI can be used for a large number of applications in a very wide range of industries—almost anywhere that two organizations, each with its own computer system, have dealings with one another.

Let us take three examples of current EDI users. These are all drawn from real life and illustrate some of the benefits and problems of working with this technology.

1.1.1 Case Study: The Acme Transport Company

The Acme Transport Company delivers goods to shops on behalf of several retail chains. Figure 1.1 is a flow chart of Acme's business operations.

Before Acme Transport incorporated EDI, each morning they would receive a pile of different instructions from each of their retailer clients. Retailer A might require delivery of two cases to one branch, three cases to another, and five cases to a third.

Retailer B would need to transport five dozen units to its branch in one city, 3 dozen to another, and so on. Clerical staff input all of these orders into a computer system that then used to work out a loading plan and delivery schedule for each of Acme's trucks. There were always some telephone calls to check details of an order, and some deliveries would often have to be held over to the following day.

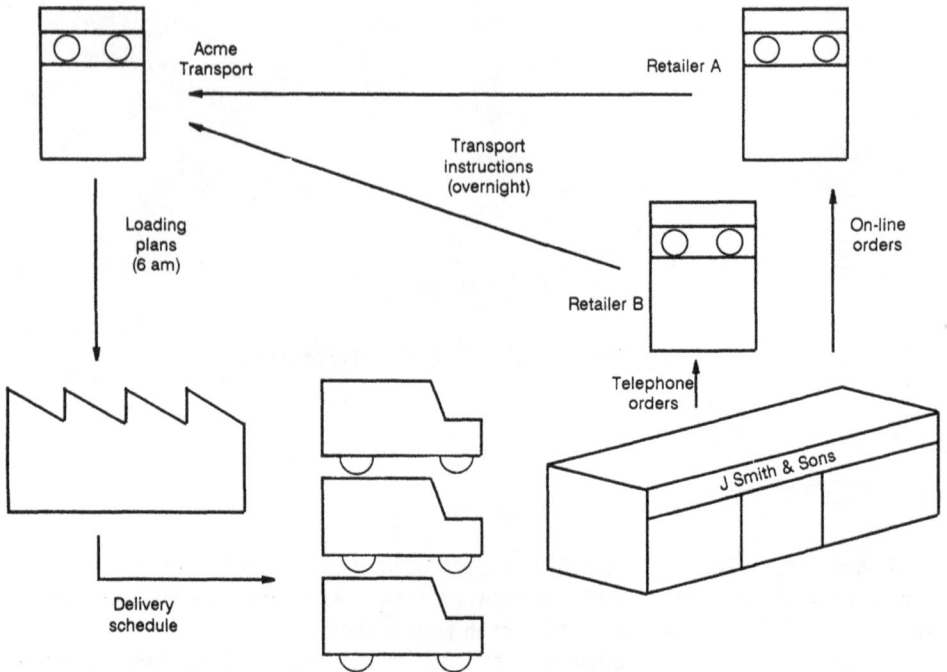

Figure 1.1 The Acme Transport Company's business operations.

To cut down on all of the clerical work, Acme agreed with several of its largest users to exchange data by EDI. Retailer A's central computer system receives data each evening from its shops; from this it calculates the stock required, and transmits a list of orders to a central EDI bureau. Retailer B receives his stock orders from the shops by telephone during the day; these are typed in and also sent in the evening to the EDI bureau.

Acme's computer calls the EDI bureau system at 5 A.M., and fetches all the orders sent by the different retailers. It then prepares the loading plans so that they are ready when the drivers report in at 6 A.M., and in many cases the goods can be in the shops by the time they open. Acme is able to give its customers a better service and also has much less wasted time for its trucks and drivers.

1.1.2 Case Study: Super Elektronik AB

Super Elektronik AB (SElAB) is a Swedish electronics-design company. All of its designs are manufactured by a small number of subcontractors who have the specialized manufacturing and test facilities required. Figure 1.2 presents a flow chart of their operations.

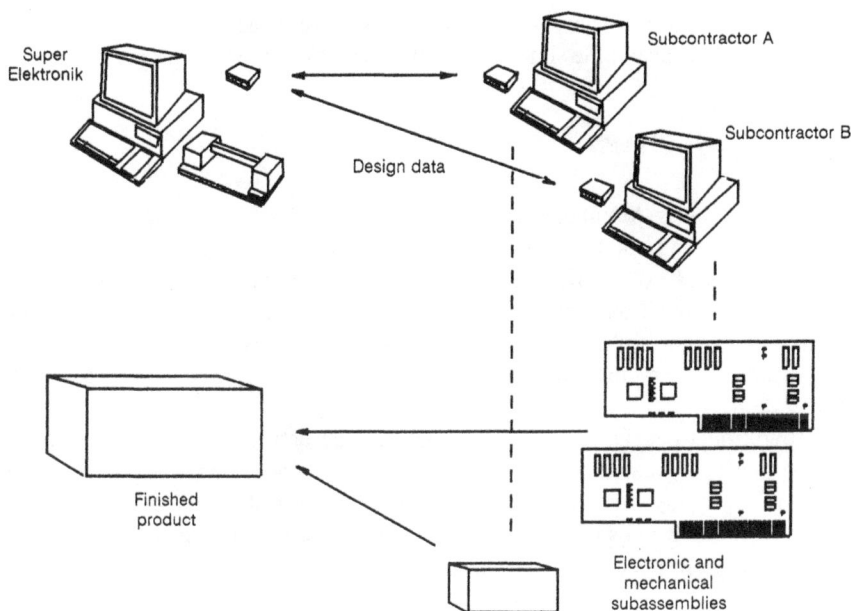

Figure 1.2 Super Elektronik AB's business operations.

A typical design involves a circuit diagram, a component list, and instructions for setting up and testing the manufactured units. With the help of a computer-aided design (CAD) program, the circuit diagram is converted into a set of masks for printed circuit-board manufacture. During this process, there is usually some exchange of ideas between SElAB and its subcontractors.

During the manufacture and test stages, several changes may be made to the design. When these were exchanged on paper, it would often take days for alterations to be fed through into the product. Copying errors were common, and since both companies were executing changes, the drawings could easily get out of step.

On one particular project, a misunderstanding of this nature caused a very expensive mistake. SElAB therefore decided to exchange data wherever possible by computer files. This imposed a certain discipline, and it became fairly easy to check that both systems were using the same version of a file. SElAB chose to use the public digital network for its EDI transfers, avoiding the need for expensive fixed telephone links between the companies.

Today, there still remains one problem: SElAB has a number of different subcontractors, each of which uses a different CAD system. Although some of the data can be ''imported'' from the one system to another, there is a real danger of confusion, particularly where a single change affects several products, each made by different subcontractors.

Therefore, SElAB is now looking at EDI standards that can be used across a wide range of CAD programs. Many of the more expensive programs can exchange data in this way; however, most of the simpler systems are restricted to their proprietary standards.

1.1.3 Case Study: CHAPS

The Clearing Houses Automated Payment Service (CHAPS) was set up by the major commercial banks in London in 1984, and offers same-day clearing for individual transactions of more than £5,000 (this limit will shortly be lowered to £1,000). This means that a company can make a payment to a business partner by computer, and the money will be credited to the partner's account on the same day. Nowadays, many banks offer their larger customers the facility to make such requests electronically. Figure 1.3 illustrates their operations.

If traditional checks are used for payment, the actual "value date" (the date on which the money is treated as cleared on the recipient's account) is variable, and can be as much as seven days after the original payment date (in the United States, often even longer). Where large sums of money are involved, these few days can mean high interest charges or delays in making further payments.

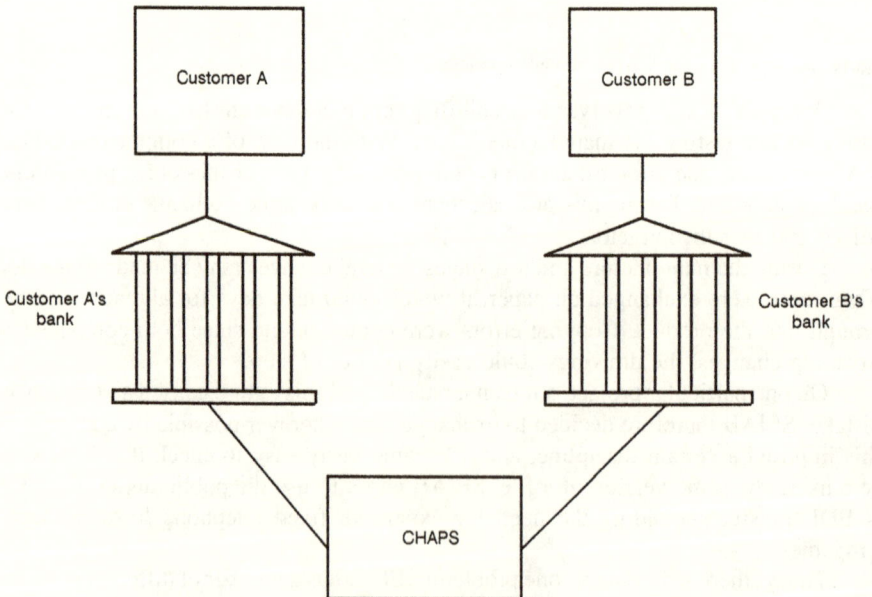

Figure 1.3 Clearing Houses Automated Payment Service.

CHAPS not only gives faster service, but the delay is fixed and guaranteed. The payment leaves the one account and arrives at the other always on the same day. There is also none of the postal delay which would be incurred with a standard check.

CHAPS is one of many EDI services that operate between banks world-wide. Massive precautions have to be taken to prevent unauthorized transactions through such a system. These involve several levels, ranging from organizational controls on physical access to terminals, multilevel passwords, and manager authorization to technical measures including encryption and message authentication.

1.1.4 Summary

These three examples not only illustrate the concept of EDI, they give some typical instances of the advantages to be gained by EDI. They also illustrate some of its continuing problems.

1.2 ADVANTAGES AND DISADVANTAGES OF EDI

The advantages of EDI include:

- EDI saves both time and manpower by avoiding the need to rekey data.
- EDI eliminates the errors introduced by rekeying.
- The data arrives much faster than it could by mail, and there is an automatic acknowledgement.
- EDI imposes a fairly strict discipline on its users. For example, it may force suppliers to quote part numbers or serial numbers or to supply certificates of delivery where these are required.

The last of these advantages is also for some people the first disadvantage of the EDI approach:

- EDI is a structured way of working: the company usually must change operating procedures.
- Responsibilities may also change when an EDI system is introduced. Unless the introduction of the EDI system and the links with other systems are managed carefully, it is possible for the data-processing department to become involved in production and purchasing decisions.
- EDI can be less transparent than paper-based systems. Under the pre-EDI system, a sales administration clerk would know whether or not Universal Widgets had placed an order on a given morning—with EDI, someone may have to check on a screen or a daily report. Under the old system, it was easy to give Universal's order top priority by putting it on top of the pile. Some companies may have to build into their system a way of allocating and changing priorities.

- Some EDI systems are very flexible, others are very simple to implement. There is often conflict between these two aims: the simplest system is usually the most rigid. For example, a really simple ordering system will include a part number, the quantity required, and a delivery date. A more typical order might be: "Hello, Bill. Can we have another batch just like the last ones you made for us, only in green. We need ten in a hurry, but can wait a week or two for the rest. And can you make sure that the outside of the box is marked for my attention?"
- Many users have already developed systems to take advantage of the fax machine, and thus avoid postal delays. Messages can be delivered by fax as fast as by EDI, and there is a form of acknowledgement. With fax cards available for computers, and with software systems that generate the text of a fax or even reconvert text into data, some users feel that they have little to gain from EDI.

Technically, there can be some problems as well:

- What happens when telephone lines go down or computers fail? How easy is it to change to a backup system? Can anyone remember how the backup works?
- Are there standards for the type of operation we envisage? If it is a straightforward retail or distribution task, the answer is almost certainly yes. But as in Super Elektronik's case, there are often gaps in more specialized areas.
- How secure is the operation? How can we be sure that our orders will reach our suppliers, and not be intercepted by competitors who may use the same service? Security is a particularly important issue for financial EDI systems, which are used for almost all interbank transfers and increasingly for communications between banks and their customers.

1.3 THE COST EQUATION

EDI involves costs. The most important of these are:

- Implementation costs. The purchase of hardware and software, installation of telephone lines and communications equipment, training and management time.
- Running costs. Rental of telephone lines, call or packet charges, other communications costs, equipment maintenance, help desks, and continuous staff training.
- Other reorganization costs. From new forms and stationery to office moves and redundancy payments.
- Study and decision-making. One of the most important and often the most hidden costs is that of researching the market, deciding whether or not to use EDI, and selecting a system appropriate to the needs of the organization. Although consultants can undoubtedly help here, they too represent a cost.

The cost-benefit equation is set out in Table 1.1.

Table 1.1
Costs and Benefits of EDI

Savings	Costs
Postage and paper	Initial capital cost
Inventory reductions	Maintenance
Better utilization of resources	Communications
(through reduced delays)	Training
Reduced errors	Consultancy
Lower transaction costs	
(hence smaller economic order quantities)	
Possible manpower savings	
Telephone costs	

1.4 "SOFT" BENEFITS

There are other benefits arising from more efficient operation, direct links with customers, and better management controls. These cannot be classed as savings, but should nevertheless affect an organization's decision as to whether or not to adopt EDI.

EDI does not automatically make a user's operation more efficient. In fact, it could make life more difficult if the manual system still runs alongside the EDI system, since the two will often conflict with one another. When implementing an EDI system, it is absolutely essential to:

- Review all parts of the operation concerned and all paperwork and reports generated;
- Decide what actually controls the operation;
- Get rid of any unnecessary operations, reports, and paperwork.

In a large organization, there may also be a potential for manpower savings. More often, EDI will allow existing staff to handle a greater throughput or to make fewer mistakes. EDI needs the full support of operational staff. It cannot be implemented in an atmosphere of mistrust, where staff feel that the system may one day make their own jobs redundant.

The telephone lines that link supplier and customer may also perform a more symbolic role. By investing in systems that both parties can use, many companies signal their long-term involvement with a trading partner. In some industries, such as banking or automobile manufacturing, it is common for an EDI user to help its customers or suppliers with the costs of equipment or software.

EDI partners are also in a position to react better to their customers' changing needs. Because transaction costs are lower, customers order in smaller quantities. Subsequently, changes in demand affect order volume and frequency more quickly, so a supplier is more aware of what is happening at the customer's end of the business. Customers are less likely to build up stocks of slow-moving goods.

1.5 DEFINITION OF TERMS

As with any computer system, EDI necessarily involves a certain amount of jargon. As this book is intended for readers from other disciplines who have to work with EDI as well as the technicians, we will try to keep the jargon to a minimum. Nevertheless, there are several terms we will use often, and it is probably as well to establish now what we mean by them.

The EDI model, illustrated in Figure 1.4, involves trading partners. These are the organizations who want to exchange data; they may be customer and supplier, two companies with a common customer, or two banks whose customers want to deal with one another. We can even have EDI between two divisions of the same company, although if they have an integrated computer system, this would not normally be treated as EDI.

The flow of data between trading partners is divided into exchanges. The simplest (and most common) form of exchange is where one partner wants to send a single message to the other, and to know whether or not the other received it. but there may be subsequent messages, in some way linked to the first, that form part of the same exchange. You can think of this as a telephone conversation carried on by telephone answering machine, between two people who are never in when the other calls.

Each exchange is divided into messages. It is on the message that most EDI standards concentrate. If we can pass messages successfully and reliably from one partner to the other, then we can operate EDI.

Each message has an originator who sends the message, and a destination, which is the other trading partner. It is possible for one message to have several destinations, although usually the protocol of the EDI system will not allow this.

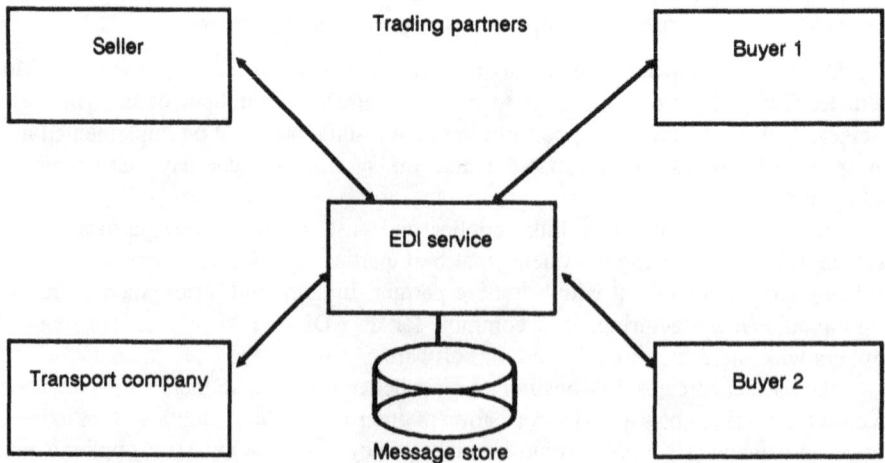

Figure 1.4 The EDI model.

The protocol is the set of rules which lays down how messages are structured, what data must be present and in what order, and what additional information can be given if wanted. It will also specify limits on, for example, the character set and the encryption methods allowed.

The protocol is part of an EDI standard. Other parts of the standard will define the electrical interface (how many wires carry which signals, what type of plug may be used), and how identification numbers and passwords for originators and destinations are allocated. The most important EDI standards for mainstream commercial use in Europe are the GTDI/EDIFACT set, which are described in some detail in Chapter 7. In North America the equivalent set is known as ANSI X.12.

The following terms and definitions should give a working vocabulary to those unfamiliar with computers.

1.5.1 Mainframe/Minicomputer/Personal Computer/Workstation/Server

These are all different forms of computer. A mainframe is a large central system, sometimes performing one giant processing task and producing an enormous amount of paper, but nowadays more often supporting a large number of users working at terminals (''interactive users'') or a network of PCs. Mainframes require special rooms to provide a clean and cool environment.

The term minicomputer is used less often these days (IBM and some other manufacturers prefer ''mid-range''), but is still useful to classify systems that conform to a number of common standards and can support a moderate number of interactive users—perhaps 10 to 40. Most minicomputers are about the size of a filing cabinet (although the actual computer part is no bigger than a PC—most of the space is taken up by the connections).

A personal computer sits on or alongside a desk, and most often is used by one person to help him or her with his or her work. The term ''PC'' is actually an IBM trademark. It is most often used to describe personal computers using one of the Intel family of microprocessors; these are described as ''IBM compatible.'' Personal computers from makers such as Apple use different standards, but are designed for the same job.

A workstation is a specialized form of personal computer or intelligent terminal. It has special features, such as high-speed graphics processing, that make it particularly suitable for certain tasks. The most common workstations today are for engineering and design, or for publishing and layout. Financial workstations are also appearing.

Many PC users need to access some common information, such as a database of customers or products. They can be connected together using a local area network (LAN; see Section 1.5.5), in which case the database will probably be held on a separate PC with a fast processor and disk—a file server.

1.5.2 Serial/Parallel/Synchronous/Asynchronous

These are all terms which describe how data is sent from to or from a computer. In serial transmission, data is sent along one pair of wires, with every ''bit'' of data following another. In a parallel transmission, all the bits of one word are sent at the same time, followed by the next word. In synchronous communications, everything is kept strictly in time by clock pulses, whereas asynchronous communications allow some variation in how fast the data is sent. You may also meet terms like 3270 and 5250, which are synchronous protocols used by IBM for communicating with terminals, and RS232, which is the most common standard for serial asynchronous connections.

1.5.3 Batch/Online

Strictly speaking, systems are online if there is a direct connection (a wire or telephone link) between a user and the computer at the time when the user's transaction is processed. In practice, the term is often used wherever transactions are processed one at a time as they come in. Batch systems process a large number of similar transactions at the same time.

1.5.4 Modem/Leased Line/Dial-up Line

As matters stand today, telephone connections to computer systems normally require a modem—a box that converts the data into sounds that can be passed down a telephone line. Lines between two fixed locations can be leased from the telephone company or PTT, or the link can be made by dialing through the normal public switched network (dial-up). The modems for each type of line are very similar, although it is possible to specify a higher speed from leased lines than from dial-up lines.

As we will see later, much of this could change when we start to use digital networks such as ISDN (*Numéris* in France), and as telecommunications deregulation allows more network operators to offer data services.

1.5.5 Network/LAN/WAN/Node

A network, in this context, is a group of computers linked together. It may sometimes include terminals, linked to the network by a terminal server. The term local area network (LAN) is used when the network is completely contained within a limited area—usually a user's site or building—whereas a wide area network (WAN) uses telecommunications lines or an external network operator to provide the connection between two or more sites.

A node is a point where the network is connected to a device (usually a computer, switch, or telephone exchange). For a wide area network, a node is the point where an external user can connect into the network.

1.5.6 Analog/Digital/Packet/PAD

An analog signal is one that can have any value within a range, whereas a digital signal can only be off or on. Digital signals are much less likely to be altered when they are passed between two devices, and this is one reason why they are used almost exclusively by computers.

Converting sounds into digital signals also makes for more accurate transmission—hence the compact disc and the move towards digital telephony. Digital networks do not require modems—data is transmitted directly.

On shared systems such as the national network, one fiber optic cable carries many signals simultaneously. For analog signals, this is done by shifting them up the frequency range by different amounts. With digital signals, it is more efficient to divide the messages into packets, each with an address on the front—rather like several letters, each sent in its own envelope, one after the other.

This requires a packet assembler-disassembler (PAD) located where the data stream is connected to the network (usually at the network node or telephone exchange).

1.5.7 Encryption/Keys/DES/RSA

To encrypt information so that unauthorized eyes cannot read it, computers divide the information into blocks and perform a mathematical manipulation on each block, using a sequence of data known as the encryption key. A further manipulation uses the data encryption key and reverses the process, making the data readable again. The forms of manipulation are often published, so whoever has the decryption key can read the data. Key management therefore is an important part of any encryption process.

DES (data encryption standard) and the RSA (Rivest-Shamir-Adelmann) method are the two data encryption standards most commonly used in EDI systems. The DES standard was developed by IBM for the U.S. Department of Defense, and was subsequently published as a standard; the same key is used to encrypt and to decipher the message. The RSA method was developed by a group of mathematicians who believed that it would not be possible to devise a code that could be deciphered using a public key without giving away the encryption key, and then proved themselves wrong. The relative merits of the two methods are discussed in more detail in Chapter 4.

Chapter 2
Opportunities and Problems

Despite all of the advantages it offers users, EDI so far has not offered sufficient opportunities for large numbers of software and service suppliers to enter the market. The number of suppliers has remained small from the start, and they are nearly all large network companies that probably make more money from providing the network services than from any of the value-added aspects of the EDI service. Why is this? Let us start by looking at the advantages of EDI, as seen by its users.

2.1 BENEFITS OF EDI

Some of the benefits of EDI are available to all users. The main ones are:

- *Faster transaction turn-around.* Suppliers do not have to wait days for an order to come through or to receive payment. This not only means earning more interest on the money (or paying less), but often simply makes the whole business run more smoothly. In many businesses, being able to react more quickly is one of the keys to competitive success.
- *Less paperwork.* The cost of paperwork involved in international transactions is estimated by the International Chamber of Commerce at around 10% of goods value. EDI can easily cut this cost in half. Although many EDI users still print out their own documents for internal use, they do not have to make copies of everything they send out. EDI should allow users to avoid filling in different forms for each supplier or carrier.

    The extent of the savings here depends on how fully EDI is integrated into the whole operation of the business. Where a company is largely computerized, and sends information around internally using computer terminals or a network, the effects of EDI reach all the way to each of the user terminals. As we will see later, this can happen despite the fact that most EDI software only updates a single master file.

- *Savings in staff time.* EDI takes on many automatic or repetitive actions: placing and calling off orders, or filling in progress forms and delivery instructions. In particular, it cuts down dramatically the number of times that the same information has to be copied or typed.

- *Fewer errors.* Operators performing typical keying or data-entry operations make errors in around two percent of entries. Where the data includes unfamiliar or highly repetitive entries, such as foreign words or columns of figures, the error rate can be much higher.

  Some of these errors can be detected by software, using range checks, check digits, or by forcing some redundant data entry. But very often, the only way to reduce errors significantly is to have two operators enter everything, and to compare the two.

  Of course, if someone makes a keying error at the beginning of an EDI operation, the wrong data are used by everyone involved. So it is particularly important to have checks on any manual entry here; nevertheless, the EDI operation reduces the amount of simple copying. Copying is more prone to errors than, for example, entering the number of cans on a shelf. With EDI, long numbers, such as part numbers, product codes or, depot numbers, that often mean nothing to the person copying them onto an order form, should never have to be keyed in for each order—they are stored on the system and used when needed.

Some of the other benefits of EDI are much more one-sided, for example:

- *Reduced inventory.* A retailer can use EDI to reduce the number of days' worth of inventory he needs to hold. If he or she previously placed a weekly order based on a stock-check the previous evening, the retailer may now be able to order on a daily basis. He or she can now keep smaller safety margins if the order is sent using EDI at the last possible moment, based on up-to-the-minute information.

  In addition, if the manufacturer is able to make supplementary deliveries during the week, even if only in exceptional cases, then he can further reduce his overstock safety margin by ordering, say, his average usage plus one standard deviation, rather than the usual average plus two SDs.

  The reverse of this applies, however, to the manufacturer. It must now keep the extra inventory that the retailer is "saving." Provided that it supplies a large number of retailers and that their demand for the product does not always go up or down together, the manufacturer's increase in inventory will be less than the decrease at the retailers, but it must still increase.

  What the manufacturer may try to do is to pass the increase on to its suppliers. If it can manufacture quickly, or increase production on demand, then it will certainly keep more of the extra inventory in the form of raw materials rather than manufactured goods. If its suppliers are local and can react quickly to order changes, then the manufacturer may try to persuade them to adopt EDI also, thus passing the effects on down the chain.

How well the manufacturer is likely to succeed will depend largely on the industry and the relative power of the buyer and seller in each case. As we will see later, though, this relationship, and the ''chain'' effect, are the keys to the growth of EDI.

- *Reduced payment periods.* Again, if one side is paid earlier, the other must lose interest. But different types of business often have different capital structures, and because of this they may have quite different costs of capital.

  A supplier who is paying interest at 18% on his bank borrowings can afford to give generous discounts for prompt payment to a retailer that is generally paid in cash. Or the retailer will simply use his payment terms to bargain for a better price.

- *Improved service to customers.* EDI can improve customer service due to the faster turn-around, more accurate invoicing and possibly more flexible operation. Here it looks as though the seller is able to gain more from using EDI. In general, though, the ''chain'' that introduces EDI into a sector passes from buyer to seller. So the advantage that the supplier gains over his competitors by introducing EDI may be short-lived.

- *Closer contact with customers and distributors.* Suppliers who make it easier for their customers to exchange information with them are likely to find other benefits from the cooperation. The contracts that enable an EDI operation generally encourage a longer-term, more stable relationship between buyer and seller. This can still be a two-edged sword: retailers, for example, can tie manufacturers into contracts by forcing them to invest in EDI technology. In some industries where the standards are not well developed, manufacturers who want to work with several retailers may have to invest in several forms of EDI.

You will have recognized a theme running through several of these lop-sided advantages: the relative power of buyer and seller. In many sectors, even where there is no monopoly or monopsony (one very powerful buyer), one of the two sides usually dictates the terms of business. In food retailing, it is the supermarkets who dictate terms; in paint, the paint manufacturers.

Where the dominant trading partner makes up its mind to use EDI (probably because it has a high degree of computerization internally), it first recommends to its suppliers or customers that they should adopt it. Shortly afterwards, the dominant trader alters its trading terms to give a significant advantage to those who do adopt EDI, and after a period of time it may even refuse to do business with those who do not. There is a big advantage to an EDI user if all its customers or all its suppliers also use EDI, since it does not have to run a manual system in parallel.

Where all or most of the powerful parties in a sector have opted for EDI, it may become an essential condition for doing business in that sector. The motor industry is rapidly approaching this state. Already, Ford of North America requires all its suppliers to use EDI—it cannot be long before other major manufacturers follow suit.

All the large companies in the European and North American chemical businesses make EDI a condition of any regular supplies, although some more specialized businesses can escape this net. EDI and a specialized labeling system are both compulsory for any supplier to NATO or to the U.S. Department of Defense.

However, joining one of the big EDI networks can be very expensive, and may present too great an overhead for a small supplier—perhaps an importer trying to sell to supermarkets.

Also, established suppliers particularly resent being forced to make this investment as a condition of doing business.

2.2 COST OF IMPLEMENTATION

The various categories of costs involved in implementing EDI are shown in Figure 2.1. The actual amounts will vary greatly from project to project, and in some businesses there may be additional costs directly or indirectly associated with introducing EDI.

2.2.1 Hardware and Software

It is possible to access an EDI system using the most basic PC system, together with some simple software provided by the EDI bureau. This method works well for small, essentially closed systems using proprietary protocols (developed by a service provider or industry group), or perhaps a small subset of the published standards.

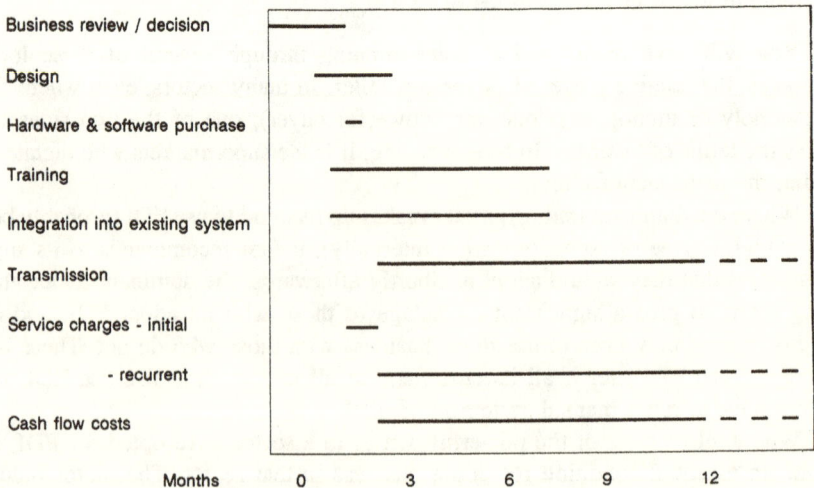

Figure 2.1 Costs of implementing EDI (in time spent).

To take advantage of the full flexibility and power of EDI, you need an EDI software package suitable for your computer system and the network you intend to use. Multipartner versions of IBM's expEDIte package or the Intercept software used in the INS/TRADA-NET system can cost several thousand dollars. A format translation package will usually be necessary, and the largest expense will often be the cost of changes to your internal computer system.

As more and more companies use 80386- or 80486-based PCs or PC networks for all their computing, the EDI software companies are "downsizing" their software to run on these platforms. There are already twenty or more packages available for PC users wishing to communicate with EDI networks. Most of these are limited to working with one network—the main ones in Europe are International Network Services (INS), General Electric Information Service (GEIS—run by GE subsidiary EDS), AT&T Istel and IBM Information Network.

PC packages are generally priced between $3,000 and $8,000; as a result they are much more accessible to the mainstream of retailers, manufacturers, and importers.

The number of software houses offering EDI software has increased considerably in the 1990s. What used to be a part of only very large retail management packages is now a significant business activity on its own. EDI software is no longer tied to any one sector.

For some of these new suppliers, support is also a major part of the business. An EDI package may have a low base price, but the software house will make its money from charges for adding supplementary message types, network or communications protocols, or trading partners.

2.2.2 Training and Initial Costs

For many larger organizations, the cost of training and introducing staff to the new system may be as large as the cost of the equipment. New forms may be needed, new data input screens, and possibly new computer-generated reports to replace existing paper reports.

Although it should be possible to simplify procedures and to eliminate some purely administrative or clerical tasks, the procedures to be followed with EDI are very important. It is quite likely that some aspects of the company's contracts with its customers or suppliers will change. Staff should by no means be taken for granted in the process—they must be aware of the implications of these changes.

Because EDI helps to reduce errors and takes away many of the more boring parts of the administration of a company, it is generally welcomed by most employees. There are always some "technophobes" who find the change to a computer representation difficult to handle. It is only fair to point out that such technophobia is most often caused by bad design of the computer system—screens can be difficult to read, keyboards difficult to use, and the consequences of making an error embarrassing.

As with any computer project, it is important to involve all staff at an early stage in the implementation. Most people will support a system when they have been involved in its design.

2.2.3 Transmission Costs

Several different communications paths may be used by EDI. Each has its own cost structure, and places different constraints on the user; they will be discussed in more detail in Chapter 5. They include:

- Conventional dial-up telephone lines;
- Direct leased lines;
- Local multiplexers, serving leased line connections to a digital network;
- Packet switched networks;
- ISDN and other high-bandwidth services;
- Satellites and other wireless communications.

One of the advantages of EDI is that users can choose when to transmit messages. It is therefore quite normal for most of the transmissions on a network to take place at night, when batch processes are most often scheduled.

Depending on the communications path used, transmitting at night may also reduce transmission costs dramatically—dial-up lines are always cheaper outside normal business hours, and many telecommunications operators offer a special tariff for large data transmissions at night.

Even when these factors are taken into account, the communications costs of a large EDI operation are a significant part of its total running costs—perhaps as much as 25% if the cost of financing the equipment is excluded.

2.2.4 Service Charges

The charge made by the EDI service company will vary considerably according to the type of service required. For full connection to a service such as TRADANET or IBM Information Exchange, there is usually a one-time entry fee, followed by monthly charges, and traffic charges on top.

At the other end of the scale, a company may provide an EDI service to its trading partners for nothing. Smaller commercial service providers charge $1,000 to $5,000 for an initial connection, $100 to $500 a year per site connected, and a message charge of a few cents. This would, however, only give access to a limited number of trading partners.

2.2.5 Cash Flow Costs

As we mentioned earlier, EDI can have a negative effect on cash flow for one of the partners (the other partner gains). This is one of several reasons why it is important to have a contract between the partners that specifically takes into account the fact that they are operating through EDI.

It may be necessary to change prices to account for this; however, when EDI is looked at as a whole, it should have the effect of lowering costs, not increasing them.

2.2.6 Integrating EDI Into Existing Systems and Procedures

This is the trickiest part of implementing EDI in any established company.

We need to start with a review of all present operations: what do we do and why? This needs to cover not only those parts of the operation that are affected directly by the EDI implementation, but also many of the administrative functions that provide support— even down to ordering stationery.

EDI messages may trigger a large number of processes, some of which will require additional information from other computer systems or manual sources. It has a particularly large effect on manufacturing resource planning (MRP2) systems in manufacturing companies, and on ordering systems in the retail trade.

Both of these are particularly complicated systems that often incorporate a large number of check procedures—documents to be completed or signatured by several managers. However tempting it may be, it is important not just to remove these checks in order to allow an EDI operation. They must be looked at critically and, if they are desirable, a way of maintaining them must be provided.

When schools in England moved from ordering all school equipment and supplies through their county councils to buying direct, most of them joined a common purchasing scheme that negotiated contracts with suppliers and set up an EDI operation between the schools and the suppliers. Whereas previously the county council had checked even every stationery order carefully, so that no school could have more than its fair share, it was now necessary to have both the head teacher and the deputy head check each order, until sufficient data could be built up to allow a consistency check to be performed in software.

2.2.7 Software Maintenance

It is inherent in the nature of EDI that its use develops over time: customers find more ways of taking advantage of EDI, take on new trading partners and new messages are needed.

The software houses that write EDI packages see this as an opportunity. Software maintenance can be a significant part of their business. For users, cost is only a part of the problem: software maintenance contracts bring with them a dependence on a particular software house.

This may be incompatible with the way the rest of the company's software has been written and maintained. The alternative is to write the software in-house, or to insist on a basic package that gives the user a high degree of control over message formats and addresses. For many users, the cost of either of these options would be too high, and they must carry the risk. In such a case, they should make sure to find a large and stable supplier.

2.3 LEGAL ISSUES

It is well-known that lawyers prefer a signed document to support any deal or contract. In many countries, electronic "documents" are not accepted as proof of a contract, and they are sometimes not even admitted as evidence.

The main objection is that it is difficult to prove that an electronic document has not been altered after it was sent. It may also be necessary to prove that it was indeed sent, or who sent it. If a buyer claims that he did not order the goods, the burden of proof is on the seller to show that he did indeed receive the order, and that it could only have come from the buyer.

There are, in EDI, two ways of countering this problem. The first, essentially an "open" route, is to have the EDI company keep a record or audit trail of all transactions passing through the network. This is a fairly onerous task, one that EDI companies try to avoid. But, using such a system, they are able to establish whether a "buy" message did or did not come from a port identifying itself as the buyer at a given time, and whether and when it was collected by one of the seller's terminals.

But what if an unauthorized person had used the buyer's password to place the message? Where security is an issue, a good password system is necessary, and users need to agree on and to use an encryption procedure. As we will see later, an asymmetrical encryption algorithm such as RSA allows the person receiving a message to check that it has been sent by the right person.

In all cases, however, the really essential thing is to have a contract that specifies in advance not only the terms of the contract, but also a mutual agreement between both parties to accept an electronic document as firm proof of the transaction, and exactly how it will be checked in cases of doubt. This is called the "interchange agreement," and is discussed further in Chapter 8.

By signing the interchange agreement, each party implicitly accepts the other's control procedures and software. Both participants should be aware of this. It is easier if all parties are using the same software and systems.

2.4 NETWORK AND DATA SECURITY

Many companies are rightly worried about connecting themselves to a network, particularly a large dial-up network that may include not only their competitors, but also other types of organization (such as universities) that are known to harbor skilled "hackers" and possibly even computer crooks. EDI is less dangerous than many forms of network connection, because it normally depends on users making the connection themselves. Outsiders are not necessarily able to dial into the system, through the network or otherwise, and most companies arrange the system so that they cannot.

There is a very small risk of a person being able to take control of a system while it is logged onto the network, but an operator would normally notice this immediately, and any good software will prevent it from happening.

Another aspect of security is data security and privacy: How can I ensure that no-one else can read the contents of the transaction or alter it? This is an important part of the design of any EDI software, and the major EDI suppliers have built extensive and clever facilities into their software to guard against these problems.

Where the need for security is even greater (for example, in interbank financial EDI), this is again a matter for encryption. Any EDI message can be encrypted, and software packages for doing this are readily available at affordable prices. Only a limited number of EDI networks can handle messages fully encrypted by the sender, however, and these are typically fairly specialized networks, offering few value-added services.

2.5 STANDARDS

This is perhaps the thorniest issue in the whole of EDI. EDI only works with standards, and its major advantages come from having widely accepted standards.

As time goes on, people find new ways of doing business and new aspects of their business relationships they want to entrust to EDI. The standards themselves must therefore also develop. A few years ago, nobody would have dreamed of trying to communicate every single aspect of a product specification and design using a set of computer files, but this is now an important application for EDI.

Agreeing on a set of standards for most common commercial transactions (the EDIFACT/X12 set) that are more or less uniform throughout Europe and North America has been a fairly major triumph for the standards bodies (although persuading users to migrate to these standards from their existing EDI systems continues to be a challenge). Beyond this, however, there is much less agreement. Although there are many users of EDI for mechanical and electronic design data, for example, different standards are used by different sectors and by different countries, and it will be many years before there is agreement on a set of standards to meet current requirements.

To some extent, this should not matter. EDI can and does provide a framework for passing messages from one organization to another. Any group of organizations can agree what these messages should contain and how they should be interpreted. The effort and time required to write a specification for use within a small group are much less than if the standard is to be used by everybody. Many organizations that refuse to participate in such an effort will not realize the greater benefits of EDI until a standard is agreed and published.

2.6 OPPORTUNITIES

Table 2.2 summarizes the benefits and problems we have so far identified with EDI. Some of these are simple: a reduction in errors is good for everyone. The middle group are, however, more complex issues and companies will regard them as advantages or disadvantages depending on their standpoint.

Adopting EDI does offer enormous gains to many users. It is enough to look through the "EDI Yellow Pages" to see how many companies can accept EDI transactions. As we have said, in some businesses it is a necessity.

It should be possible to extend EDI from the simple commercial transactions that are common today to cover most exchanges of information between companies, government bodies and other organizations. When EDI is linked with electronic mail, it can replace almost any letter, fax, or telex.

EDI offers its main advantages where the data are originated on a computer system, and are to be transmitted for use by another computer system. This covers an increasing amount of commercial data. There will be a need for a much larger number of networks, able to carry small numbers of large transactions as well as the large numbers of short messages that are normal today.

It looks as though the standards for design data, for example, will be very complex. Many software houses currently producing computer-aided design software will have to become involved in writing software for EDI. There will also be a need for software companies that can handle the more complex communications tasks needed by the new forms of EDI.

Table 2.2
Benefits, Problems, and Issues

Benefits	*Problems*
Faster transaction turn-around	Legal
Less paperwork	Security
Savings in staff time	The need for standards
Fewer errors	

Issues

Reduced inventory versus shift in inventory responsibility

Faster incoming payments versus faster outgoing payments

Closer links with customers versus being tied to suppliers

In Part II, we will examine the functions of an EDI system and the hardware, software, and communications systems required to implement them before looking at the many applications already in use or being planned. In Part III, we return to the opportunities and problems in EDI and try to see what the future offers for users of this exciting yet common-sense application of modern technology to everyday business problems.

Part II

Chapter 3
Functions of an EDI System

So far, we have looked at the overall function of electronic data interchange, and at the problems that it sets out to solve in various environments. Now we need to look at the way an EDI system is designed and the functions of its various component parts.

3.1 THE EDI PROCESS

The purpose of an EDI system is to pass data from one computer system to another. It does this in a number of stages. Typically, an EDI system will:

- Take the output from some process on one trading partner's computer (such as an inventory check or a design step), and determine what data needs to be sent to the other partner via the EDI system.
- Format the data into messages according to the requirements and protocols of the EDI system.
- Connect to a central computer system (the "EDI bureau"), owned by a third company, and carry out any security checks.
- Send the messages to the EDI bureau. These are then stored in a "mailbox" for the destination computer, possibly after some further format conversion.

Once stored at the EDI bureau, the rest of the process works as follows:

- The other computer calls the EDI bureau and, after carrying out similar security checks, fetches the messages.
- The messages are decoded and the data stored in a form that can be used by the next process in the chain.

Throughout this procedure, accounting processes and audit trails are maintained. Additionally, there will often be control processes in both trading partners' computers to start off the EDI functions and to pass messages to and from other parts of the computer system. Figure 3.1 is a diagram of this process.

3.2 EDI FUNCTIONS

We can look at these processes in terms of the functions the EDI system performs. Without EDI, companies trading with many partners might have to have computer links to each of their partners' computers. Even if a switched network is used, this would mean dozens of different standards, protocols and passwords. With an EDI system, all of the companies only need to have one link with the central computer, or a node on its network. EDI performs a communications function, as shown in Figure 3.2.

There can be quite a few variations on this sequence—for example, EDI formats can very well be used directly between two computers, without the need for a central system. But this removes one of the advantages of EDI, which is that the two companies can operate at their own speeds, without needing to be synchronized. EDI performs a buffering function, as in Figure 3.3.

Most EDI systems define rigid formats for the structure and content of the input data, as well as the structure and protocols for messages. This means that all format conversions must be done in the originating and and in the receiving computers. Other central systems specifically help the users by converting from the originator's format into a standard format or the receiver's preferred format. Wherever it is used, EDI must provide a format conversion function (Figure 3.4); this will usually be in the trading partners' computers.

Although in a few cases completely private networks are used, most EDI systems depend on wide area networks such as the public switched trunk network (PSTN) or a packet switched (X.25) network. It is therefore normally important to ensure that only the right people can deposit or gain access to data on the system. EDI must therefore perform a security function.

Linked to this is the need for systems sending messages to be sure that they have been received by the destination system. EDI systems keep track of the time messages were sent and received, and provide a backup in case messages were lost or systems failed while processing a message. This same data is normally used by the service provider to charge for the service. Thus, EDI also provides an audit function.

Further records are kept so that the EDI service provider can charge users for the service; this will be linked to the security (access control) system so that users who have not paid can be barred from the service. The system providers also need statistics of usage in order to calculate their costs and returns. As such, EDI systems provide an accounting and analysis function. The audit and accounting functions are shown in Figure 3.5.

Let us look at each of these functions in turn and work out what the requirements are. In the next three chapters, we will see how the software provided by EDI companies, the communications networks, the hardware at the central and user sites carry out these requirements.

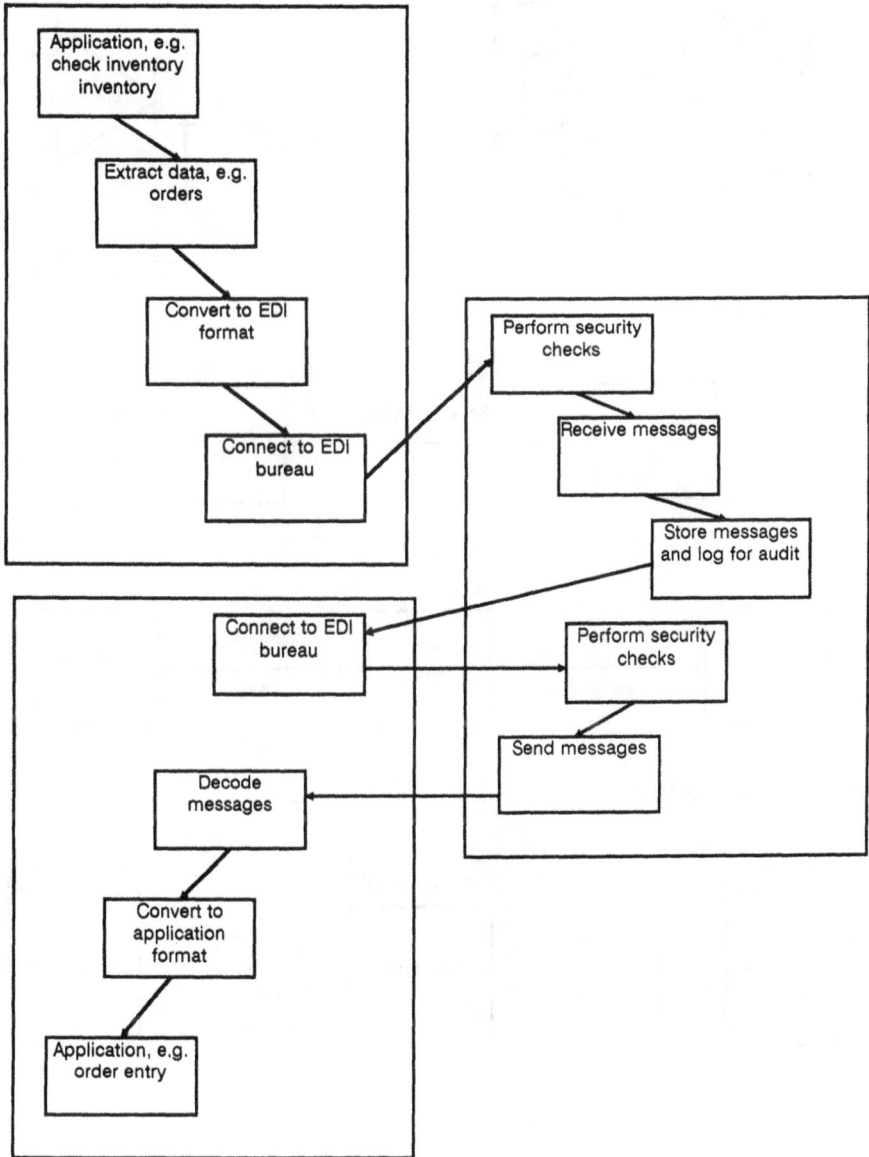

Figure 3.1 Sequence of the EDI process.

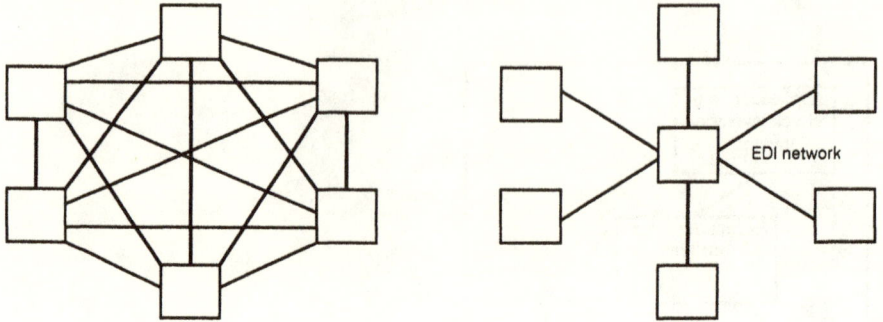

Figure 3.2 The communications function.

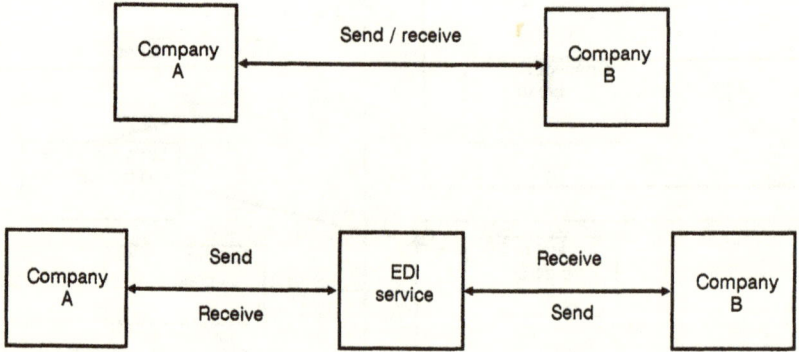

Figure 3.3 The buffering function.

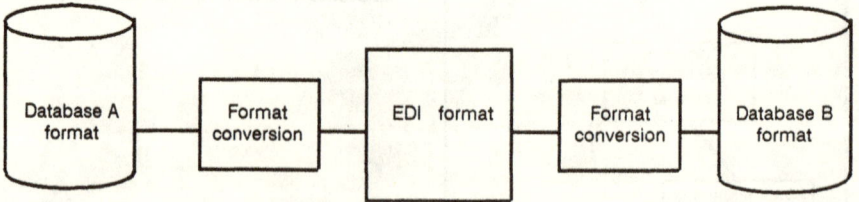

Figure 3.4 Format conversion.

3.2.1 Communications

Most EDI systems connect different companies via the central EDI computer. The users'
computer systems are not just peripherals, like a disk drive or a printer, attached to the

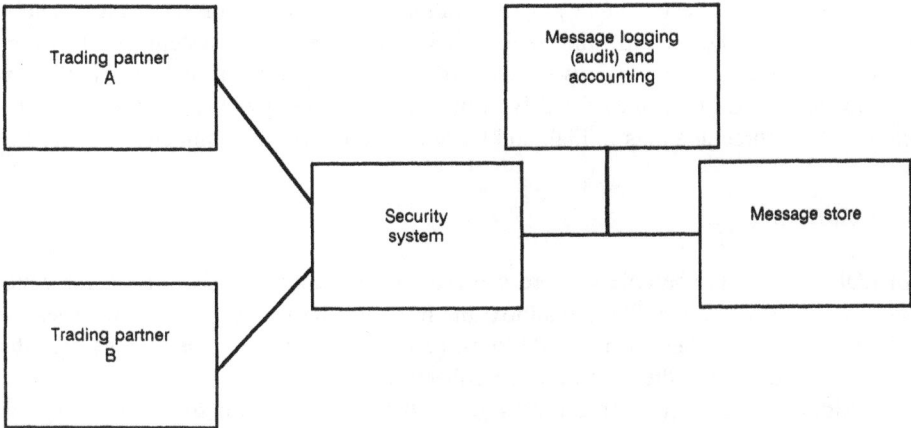

Figure 3.5 Audit and accounting functions.

central system; in fact, some of them may be much more powerful than the central system. When several computer systems are linked together like this, it is called a network.

The systems are likely to be more than a few hundred meters apart; they may be in different towns or even different countries. So this is a wide area network rather than a local area network such as you might find in a single office or factory site. WANs can connect together one or more LANs; they always used to run more slowly than LANs, because they were based on telephone cables. Now, however, WAN connections use fiber optics, so some of the differences between LANs and WANs are starting to disappear.

Most EDI exchanges are quite short: the equivalent of a few pages of text, at the most. Users call the EDI system from time to time—once a day, in many applications—rather than being permanently connected. The normal answer to this would be to use a dial-up line—the PSTN. But, as we will see in Chapter 5, there are now several alternatives to this, preferred by EDI system users in most, but not all, countries.

What we often have is an EDI network that makes use of another data communications network. Although the word network is used for both, the first is a logical network, and the second a physical network. This kind of distinction crops up quite often in computing. In this case, it means that the logical network is concerned with the information it carries (the EDI messages), while the physical network consists of boxes connected by wires or some other medium, and is only interested in passing a stream of data to the right destination, without being concerned with the messages that are contained in the data.

The communications system used must be economical for this kind of intermittent traffic; very often, it is the most expensive part of the EDI service for the operator, and it is a service that he must buy. Often, there are two charges: a public network charge that the user will have to pay, and the cost of the private network to which users connect.

It must also be very reliable; communications links have become much more reliable in the 1980s, but they are still the weakest link in most computer systems. EDI systems must accept that communications errors and failures will happen, and they must have fallback links and procedures for these times. For dial-up systems, most systems will allow at least three attempts to dial, and these may use different numbers.

3.2.2 Buffering

An EDI system must be able to store messages for collection by the system for which they are intended. This is like a mailbox: the box must be secure, so that messages can only be collected by their owner, and it must be big enough to store all the messages that anybody could send in the time between collections.

Although it does not concern most users, the design of the output mailbox system is often one of the main factors determining the design of the EDI system. It not only sets the level of security of the system, but is also one of the most important factors in making the system easy to use.

EDI exchanges often take place at night and at weekends, taking advantage of cheap communications rates. Trading partners must agree on a timetable for sending and receiving messages. In some cases this will also be controlled by the central EDI system.

3.2.3 Format Conversion

At the heart of the EDI concept are the standard formats for data items and messages.

A few EDI systems can accept "free formats;" in other words a sender may use his preferred format, and the EDI system itself will automatically convert it into the format preferred by the receiving system. This mode of operation suits small companies who expect to make only occasional use of EDI. It is, though, very much the exception. Nearly all EDI is based on formats defined in a set of standards, such as those in the American National Standards Institute's X.12 set (ANSI X.12), the European GTDI standards or the newer EDIFACT standards (EDIFACT is described in more detail in Chapter 7; it is sponsored by the United Nations Economic Commission for Europe (UN/ECE) and has been designed specifically for international use, thus bridging the gap between X.12 and GTDI).

The job of the format conversion software is therefore to take the user's data, probably created by an external application program, and to convert it into EDIFACT/ X.12 or whatever other format is being used. The same applies in reverse to the receiving system, which must be able to accept and interpret data in the EDI format.

Because of the very wide range of application packages that can generate, say, a purchase order, special software is often required. Fortunately for most users, general EDI format conversion programs, using tables of messages and formats, are available from several vendors; these are discussed in the next chapter.

In sectors where EDI is common, application software often includes an option to output or input data in the appropriate EDI format. Sectors that do this already include publishing and electronic design, while the latest generation of retail systems incorporate software "hooks" that allow users to build the format conversion functions into their systems.

It is really important for users contemplating EDI to try to ensure that their systems do have these functions. Having special software written considerably increases the cost of implementing EDI.

Format conversion is not necessary where the whole system has been designed around the EDI requirement, so that these are the only formats used in the system. This would, for example, be the case in the financial EDI packages provided by banks to their customers to allow them to make transfers electronically.

3.2.4 Security

The main aims of the security function in an EDI system are to ensure that:

- Nobody can leave a "fake" message on the system without it being easily detected;
- Nobody can alter or delete any message while it is in the system;
- No user system can receive or read messages not intended for it; and
- No activity by one user can affect any other user.

Nobody will use an EDI system that they do not consider "secure," but users have very different ideas of what constitutes "security." They normally concentrate on access control, which is only a part of a comprehensive security system.

Some security systems just make signing on to the system a long and complex process, without necessarily making it more difficult to bypass the system. In fact, the longer and more complicated the access control procedure, the more likely it is that it will be written down or fully programmed into some equipment.

Hackers have often shown that it is easier to bypass a system's security than to "break" it. They most often obtain a valid password set by asking someone for their password on some pretext, copying a set that someone has written down or programmed into a terminal in clear form, or by using engineers' passwords that have not been deleted. This last form is particularly dangerous because engineers can often use system-manager functions.

For an EDI system to be secure, it must:

- Only be accessible using special software, where the software itself cannot be copied or used to transmit data other than the EDI messages from a particular source;
- Positively check the identity of every computer logging in to the EDI system. Some access control systems simply assume that any system that gives a correct password sequence is legitimate; and
- Not use any analog telephone lines, or other system where it is easy to connect a monitoring device to the line undetected and decode the full exchange.

The computer systems that legitimately store, forward, originate, or receive the data must also be physically secure. If they are multiuser systems, they must have adequate interprocess protection. Most larger systems do have this, but many PC-based systems are weak in this area.

Highly sensitive systems, such as those used in interbank EDI, would add to this list. In overall effect, it must be impossible for any part of any message or exchange, including the access control sequence, to be read, altered, or deleted by any outside party. It must equally be impossible for any party other than a legitimate user to originate a message.

In the design of these systems, it is even necessary to recognize that signals passing along wires can be monitored by means of the minute amounts of electromagnetic energy that they radiate.

3.2.5 Audit

All EDI exchanges must be logged. As well as knowing that messages cannot be added, altered, or deleted, the EDI system needs to know that every message has been received and collected.

First, this implies a numbering system, so that missing messages can be detected. Then there must be a history file, or audit trail, that keeps a record of each session (a session is a period when a user is "logged on" or connected to the EDI system), and what messages were deposited or collected in that session.

There is no overall standard for keeping audit trails. As mentioned in Chapter 2, lawyers are taking an increasing interest in EDI, so any systems that form the basis of contracts, whether formal or implied, between the trading partners, are likely to have to conform to certain minimum standards.

It is also useful if the audit system produces exception reports—for example, where a user has an extra session or misses a session compared with his normal pattern, or where a message has not been collected or delivered within the normal period.

3.2.6 Accounting and Analysis

The accounting system is often tied in with the audit function, although (despite their names) there is not really any logical connection. It is simply that the data on the audit system is normally what the accounting system needs to prepare bills and usage analyses. The factors that are normally measured include:

- Connect time: The length of time for which users are connected to the system;
- Transactions: The number and, sometimes, length of messages sent and received;
- Storage: The amount of storage space reserved for this user. As well as any "mailboxes," this may include special data held on the system for reference purposes.

3.3 OVERALL FUNCTIONAL REQUIREMENTS

3.3.1 Reliability

The key requirement for any EDI system is its reliability. The value of the transactions being transmitted is much greater than the cost of the service, and any delay or, worse, an error in the data could lead to severe losses, embarrassment, and legal claims. International money transfers of millions of dollars pass by EDI, while a small delay in sending an order for parts in a JIT (just-in-time) manufacturing system can hold up the complete production line.

Almost more important than the actual reliability is the impression of reliability given by the system. It is not possible to make any system 100% reliable, but the way that the system handles failures, the speed with which the system recovers, the reports given to users, and the attitude of the staff are the way users measure reliability.

3.3.2 Geographical Coverage

An EDI system should be an "open" system, in that it should be possible for any trading partner of an existing user to connect to it. Since these trading partners may be very widely distributed geographically, the EDI network must also have a wide coverage.

This requirement is often met by interconnecting networks: as we will see in Chapter 15, most of the major EDI networks feature links to other value-added networks.

Retail and wholesale distribution is organized on a national basis in Europe, and on a regional basis in North America. This means that the communications network for an EDI system in this sector, for example, must have at least national coverage in Europe. Only a small handful of retailers operate across national boundaries.

In some other sectors, for example in electronic design or interbank EDI, EDI is used as much internationally as nationally. In these cases, the network used must have a very broad international coverage.

3.3.3 Connectivity

By the same token, it must be possible to connect any type of equipment to the EDI system. Historically, this was done by converting everything to the "lowest" common format: an asynchronous, analog modem.

Nowadays, networks use "intelligent hubs," "routers," and "gateways" to allow different devices to connect to the same network. Intelligent hubs and routers are used where all the devices are using the same protocols (for example, Ethernet or Token Ring); routers can also make decisions about the best path for a particular message to take. A gateway is used where the devices use different protocols. This is the type of device that would be of greatest value in an EDI network.

3.3.4 Application (Value-Added) Functions

An EDI network's main role is to pass messages in a standard format between two or more computer systems. By virtue of the way it is connected, there are actually many other functions that it can perform.

For example, an EDI system in the transport industry could also offer a system for matching loads to "empty legs"—journeys where a vehicle has no load in one direction. The messages could be passed in an EDI format, and matches notified similarly.

Or the system can perform additional processing on the data for a particular customer. In the case of Acme Transport, described in Chapter 1, the EDI company could actually do the allocation of loads to vehicles, provided that all the orders came in via EDI.

Any functions of this type would have to be specified individually, and would be highly dependent on the sector concerned. The opportunities for this kind of function will be greatest in sectors where companies have limited computer facilities, and where competition is a fairly significant factor but not extreme. If companies have plenty of spare computer capacity and manpower, they are unlikely to subcontract such value-added functions. If there is little competition, the need for economy is not so great, while in cases of extreme competition, companies will not subscribe to a service that all their competitors can also use.

Chapter 4
EDI Software

4.1 SOFTWARE FUNCTIONS

Software performs most of the EDI system functions described in Chapter 3. But it can only do so in the framework of a suitable communications network and adequate computer hardware and physical and business procedures. We will emphasize in several places that good EDI system design depends more on the procedures and the communications network than on the software itself.

Having said that, the selection of a suitable software system is clearly an important decision that may have strategic as well as practical effects on the company's EDI development program.

Later in this chapter we will talk about open systems interconnection—the OSI model. This provides a framework and a vocabulary for software and hardware designers involved in data communications, particularly where different types of hardware or software are involved.

Properly designed and implemented EDI software will avoid the need for users to become involved in the detail of the individual standards used at each level. Ideally, an EDI system should follow OSI preferred standards throughout, but in practice this is not usually the case.

The EDI user will be more concerned with the functions and features of his system. In this chapter, we consider the way that software can provide these elements.

4.1.1 Format Conversion

The job of the format conversion software is to read data output from an application package (such as the ordering requirements thrown up by an inventory check), and to convert all the data from the formats used by the in-house system into the agreed EDI

---

standard format. It then performs the reverse operation on data received from the EDI system, converting it back into the in-house format.

The operation is performed in three stages: first, the software decides which message is to be used, either from the instructions passed to it by a user at a keyboard or from a "batch file" or job control program that performs a set sequence of operations at a particular time of day or night. It then looks up the sequence of items required for that message: for example, an order might have the sequence:

ORDER
>From_____
>To_____
>Dated_____
>My reference_____
>Goods required_____
>Part no._____
>Description_____
>Quantity_____
>Required delivery date_____
>Delivery method_____

This will probably be defined in a table, with the items specified by their item codes as specified in a directory such as the UN Trade Data Elements Directory (which is discussed in more detail in Chapter 7). This table will specify not only the content (meaning) of each item, but also its length, whether it can contain letters as well as numbers and so on.

Last, the EDI software must go to the in-house software and find the necessary data, converting the lengths or codes used to those required by the standard, and adding any fixed fields or punctuation. Depending on the nature of the two software systems, the internal system may have to be modified to produce a file for the conversion software to work from; in other cases, the EDI software may be able to extract directly from a stock database.

Because both the in-house software and the standards may change from time to time, it is usually best for all the format conversion to be driven by a set of tables. It is then much easier to update them when other factors change.

When data is received, the same operation is carried out in reverse. This time, it may also be necessary to divide up the message received, as one EDI message may correspond to several entries on the user's database. The EDI software must also take this into account.

4.1.2 Message Management

Once the individual messages have been constructed by the format conversion software, the next task is to build them into a stream of data that can be passed to the communications system.

The message-management subsystem will take the messages and add the ''headers'' and ''footers''—the short sequences of data that identify the start and finish of a message—together with some control information. The subsystem will also keep a record of the messages transmitted.

It may then build up a number of these messages to form an interchange, which is the whole sequence of data sent to the EDI bureau or other partner through the communications system.

4.1.3 Communications

The communications function is the most tricky part of any EDI software system. It varies greatly from case to case, even sometimes within one EDI system, and there is often minimal help available from the software developers.

As we discuss in Chapter 5, the range of communications options available for EDI usage is very wide, although fortunately only a few of these options will be open to a typical new EDI user. The most common route, using the public switched network and external modems, is also the most troublesome.

EDI needs a system that is optimized for sending and receiving files; most common communications packages are not really suitable for this application, as they wait too often for acknowledgements, and thereby slow down the transmission by a factor of at least two. When testing communications software, you need to check that it transmits files at the speed of the modem and network; otherwise you are paying for unnecessary connect time.

Many EDI systems now make use of value-added networks. These are likely to have specific requirements for logging on and initiating transfers. It is generally best to use software that has been specially designed for that network.

Digital systems are generally easier and more reliable than analog, but are not yet widely available for general EDI use. They are most likely to be used today in private networks and in France, where the digital network is most developed. The rate of penetration is now increasingly fast in other countries as well.

In the case of packaged solutions, the communications software should already have been selected or written with these issues in mind.

4.1.4 Security and Access Control

EDI is a very powerful tool, and must be kept under careful control. EDI software should perform at least two sets of security checks: on the operator sitting at a PC or terminal and giving commands to the system and on any communication with the network or external EDI system.

As regards the operator at the terminal, the control should be physical and logical. You do not want just anyone to sit at your PC and type seemingly innocuous commands such as "del *.*" under any conditions. It is even more important that they should not be able to place an order for, say, a dozen warships or 5 tons of jelly beans.

Single password controls on their own are fairly ineffective except against casual misuse. The minimum should really be two levels of password, or a password and a token combined.

The software can also contain controls against errors, both manual and automatic. EDI systems frequently limit the maximum size of an order to a particular supplier or the maximum size that will be accepted from a particular customer. These are very important, as interchange agreements (see Chapter 8) usually require both parties to honor any correctly placed EDI instruction, even if it is erroneous. In some situations, it will be an advantage to allow a manual check of the data that is to be sent out.

Computer system users must take extreme precautions when logging on to any network. Unless it is a fully private network, and there is no possibility of anyone else physically connecting to it, the assumption must be that there could be anyone on the far end. The computers should indulge in some fairly intimate small-talk before getting down to business, with the level of intimacy dependent on the nature of the business.

Although most EDI standards allow for encryption of messages, this creates difficulties for the EDI system operator responsible for the audit trail and for recovery from errors. Some operators also monitor, or reserve the right to monitor, the content of messages to ensure that they do not break any laws.

The compromise usually adopted is to encrypt a password and some control information. The control information should include enough data to ensure that messages can be neither missed nor altered; a message counter is nearly always included, and some form of checksum (the last few digits of the sum of all the characters in the message) is also desirable. In very sensitive cases, particularly in financial applications, an encrypted form of "digital signature" may also be taken from the user's token or log-on sequence.

The Data Encryption Standard algorithm developed by IBM for the U.S. Department of Defense is commonly used in data communications; it is the easiest to implement, and is very secure provided that the encryption keys are not compromised.

In an EDI environment, however, it is very difficult to maintain secret keys: there either has to be a key for every pair of trading partners or a set of system keys available to all legitimate participants.

A much better system is offered by an asymmetrical key system developed by the mathematicians Rivest, Shamir, and Adelmann, known as the RSA algorithm. In this case, each participant can publish a key; information sent using this key can only be decoded by the key owner. Although still not perfect, this goes a long way towards meeting the functional requirements for security described in Chapter 3.

4.1.5 Housekeeping and Audit

An EDI system is usually run by a system manager; he or she has a special password that can be used for registering new users or changing their levels of access.

The system manager also needs to be able to control the sequence in which jobs are run, or to update the messages or directory in use. He or she should also be able to initiate extracts or transfers manually, perhaps when recovering from an error.

These actions, along with a comprehensive record of all messages sent and received, are recorded in the system log, which acts as an audit trail. Usually two forms of media are used. For example, if the main system runs from a normal magnetic disk, the log is stored on tape, on optical disk or even on paper (using a secure printer).

Figure 4.1 shows the structure of a typical EDI software suite.

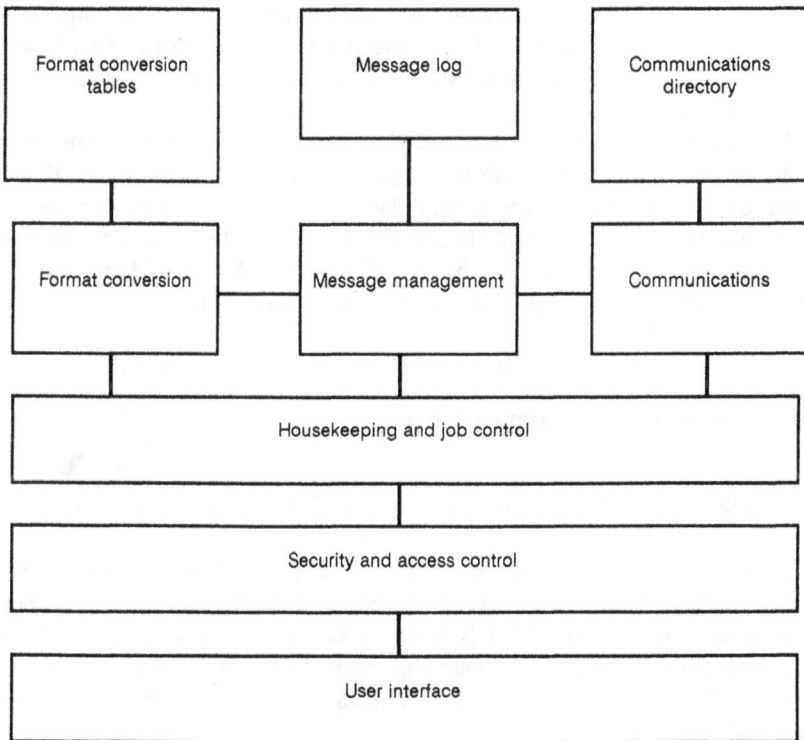

Figure 4.1 EDI software structure.

4.2 IMPORTANT FEATURES

Some new EDI users will have no choice as to the software they use. A dominant trading partner or trade association may impose (or in some cases even provide) a software package or solution. The software may be provided as part of the EDI service, so that choosing a service also selects the software. It is still best to know how features may vary from system to system.

4.2.1 Hardware Platforms

One of the most obvious areas is the range of hardware or operating systems for which a package is available. Early EDI software was written for mainframe systems, using IBM 3780 protocols. The Digital Equipment range has also been popular with many EDI system providers and with several of the large multisite operations that have been the first users of EDI. Many systems are therefore available for these systems, including Digital's own VAX/EDI.

The current trend is towards smaller EDI platforms: PCs, on their own or very often on a network, can easily cope with the requirements of most EDI users. Here, it is important to consider the operating system platforms in use. A system that can only be used under the basic MS-DOS, without any networking capability, may be adequate on day one, but may very soon restrict the growth of the EDI system. Most of the more advanced PC systems can also operate on networks or under UNIX.

4.2.2 Range of Standards Accommodated

There is a similar consideration in relation to the EDI and communications standards supported by the software.

As we will see in later chapters, there are many EDI standards in use. There is, however, a tendency to converge towards the one EDIFACT syntax, and to incorporate any new messages required into the UN Standard Message set. As a minimum, therefore, EDI software should be capable of handling the current standard used in the relevant sector and EDIFACT, and should have a mechanism for incorporating future standard messages and directories.

On the communications front, it is perhaps more difficult for a software developer to take into account all the future possibilities. Perhaps the best that can be hoped for here is a commitment to meeting or working with future standards. If the package only accommodates one type of communications today, there must be some doubt as to its flexibility to cope with future systems.

There is a general trend towards wider connectivity. For many EDI networks, this means a link into existing X.400 networks; the X.435 standard, which is now the preferred method of transmitting EDI messages over X.400 networks, should therefore be taken into account.

4.2.3 Communications

As we mentioned earlier, it is important that any EDI software is optimized for the particular network or form of communications on which it will be used. The differences in efficiency are very large.

Good facilities for handling and recovering from errors are also important. An EDI system must often work unattended, and any messages given by the system or network must be correctly interpreted and acted upon. It is surprising how many systems offer messages like "System busy: retry?"

An aspect that is ignored by many EDI packages is the number of different EDI networks now in use. Many users have to use one VAN for receiving orders and sending delivery documents, another for posting shipping documents and customs forms, and perhaps a third for electronic payments. One large customer may have its own private network, while the design company wants to be able to send and receive design data directly over the PSTN. The software must be flexible enough to route each type of message correctly.

4.2.4 Ease of Upgrading

It is important that new forms of transaction, new trading partners, or even new networks and directories, can be added simply. This has been a design requirement for most major EDI systems, which are table-driven as a consequence, but the actual ease of implementation still varies considerably. EDI software purchasers are advised to go through these procedures in some detail.

4.2.5 Ease of Use

An EDI system should run with the minimum of human intervention; after all, one of the main benefits of EDI was the elimination of both the delays and errors that this causes.

Nevertheless, there are times, as described above, when an operator should check a transmission, or initiate some action. This should be relatively easy to carry out, bearing in mind the security constraints of the system.

The interface between the EDI system and the application is also important here: it is possible for the EDI system to get in the way of the application, and hence of the business as a whole. Examples of this are when, in single-user systems, enquiries on the stock or MRP system must stop while the EDI file is being extracted.

4.2.6 Security

The way in which the software implements the security functions described earlier in the chapter is also very important from the user's point of view. Whereas in some situations an impression of security may be more important than the actual level of security offered, the EDI community today is small enough to include many astute analysts, and "cosmetic security" is unlikely to stand the test of time.

4.3 CATEGORIES OF SOFTWARE

EDI software falls into various categories.

- Generic message handlers. This is the most common form of EDI software. Message handlers perform the full range of format conversion, message management, and user interface functions, but are often weak or restricted on the communications side. Message handlers are available for all main PC and minicomputer systems and for all the major EDI networks. They are generally the preferred choice for users interfacing with well-developed in-house applications.
- Packaged user systems. This is the type of package that is sometimes provided as a part of an EDI service. Because they are designed to function with a particular system, they are often restricted in the types of communication they can support, but will be optimized for the given EDI service and will require little or no initial setup. Such packages may demand more extensive modifications to the in-house software to produce files in exactly the required format.
- EDI gateways. It is no longer difficult for a company to set up its own EDI system, including a "store and forward" facility. Such a gateway must accommodate many different networks and standards, but may be necessary where the range of EDI applications within the company is wide.

Large EDI users, including retailers and consumer goods manufacturers, often use direct links where the volume of data passing between them makes a VAN expensive.

4.4 OPEN SYSTEMS INTERCONNECTION

The technical literature on EDI (and indeed on all data communications) is littered with references to "layers" and acronyms like GOSIP and FTAM. It is not necessary to understand the details of OSI in order to understand EDI (it would be a brave person who claimed to understand all the details of OSI anyway), but a knowledge of the concepts is useful to anyone who will be involved in the technical side of EDI.

The open systems interconnection model defines seven "layers," each of which represents a set of alternative hardware or protocol standards. The lower three levels define the actual data transmission standards used, while the top three represent aspects of the user interface, with level 4, the transport layer, providing the link between the two (see Figure 4.2).

The physical link layer, level 1, represents the cable or optical fiber used for the interconnection, together with the voltage levels, speeds and connectors. The most com-

Level

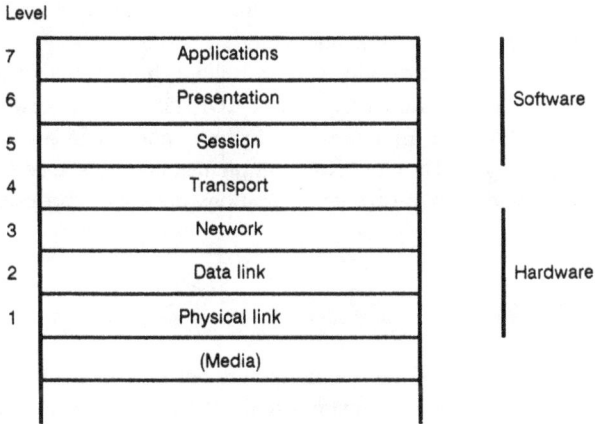

Figure 4.2 The OSI model.

monly used standards here are RS232-C serial, the various coaxial cables used by Ethernet, and fiber optics.

Level 2 is the data link layer. This determines how the data is organized into messages or packets and how the network is synchronized. The main OSI-compatible standard used in EDI is the CCITT X.25 packet switched network (Datex-P, PSS or Transpac in Europe).

The network layer, level 3, controls and manages the network. This includes many aspects of addressing and routing, as well as handling error conditions. Several of these conditions are in practice defined in level-2 standards, but complex networks have a need for additional control functions. Again, the preferred implementation is ISO 8208, which is a protocol for using X.25 in terminal devices.

The interface between the software-oriented user functions and the hardware levels just described is provided by the transport layer. This corresponds to the operating system in a computer, and is most commonly provided by the VAN itself. Although OSI was conceived with physical connections in mind, a "connectionless mode" has been defined for EDI and similar applications.

For some networks, there is little or no need for a session layer, level 5. In multiuser environments with varying access rights, it may be critical. It may be provided as an integral operating system function, add-on software, or even a separate processor.

Level 6 is the presentation layer, concerned with data formats and screen presentation. Some aspects of the presentation may be closely linked with the application itself.

The application layer, level 7, may be a self-contained communications application or an interface to an application. Examples are EDI, office document architecture (ODA), and file transfer, access, and management (FTAM). One of the best defined is the X.400 electronic messaging standard, which has given a significant boost to the use of OSI standards.

EDI links an application in the first computer to another application in the second, and should therefore be viewed as encompassing all seven levels.

Few if any EDI systems are fully OSI-compliant, however; this is partly because the basic syntax of X.400 is different from most EDI standards. A new standard, X.435, has been defined to allow the EDI message to be contained in an X.400-compatible "envelope," and this may lead to more X.400 implementations. The advantage this offers is access to the very wide range of communications networks and users already using X.400.

4.5 INTERACTIVE EDI

Mention should also be made of the special requirements of interactive EDI. Interactive transaction-based systems, where companies make their databases available to outsiders for enquiries or orders, have grown rapidly during the 1980s, particularly in the travel industry and in certain retail sectors. These have tended to use the videotex standards: Prestel in the UK, Minitel in France, Btx (CEPT-3) in Germany and Switzerland.

These standards were very easy to use, and offered inexpensive terminals. As other, particularly PC, hardware has come down in price, and faster data communications standards have come into use, they no longer meet the requirements of more intensive users.

Many users of these interactive systems also have a requirement for other functions— invoicing, insurance and the like—that are now available in one of the EDI standards. It would be an advantage to them if their interactive traffic could also be formatted as EDI messages and transmitted over the EDI network, with special routing information.

Interactive EDI systems are currently being designed and tested. A set of rules for IEDI was finalized by UN/EDIFACT in 1992.

Chapter 5

Communications Networks

5.1 CHOICE OF COMMUNICATIONS NETWORK

As we will see in Chapter 8, the choice of a suitable form of communications network is one of the first decisions to be made when implementing an EDI system. It is also a decision that must be reviewed frequently, as the pace of change in data communications is very rapid, and a user's requirements will change as the scope of his EDI operations increases.

Most EDI users, so far at least, manage with only one form of communications network, although sometimes one connection to a network can hide the variety of links and gateways that form that network. Sending disks or tapes by post or courier can also be considered as a form of EDI, since it connects a process in one computer with a process in another.

So what are the choices for the communications network to pass our EDI messages? What are their respective advantages and disadvantages and to what extent are they all available today?

5.2 PUBLIC SWITCHED NETWORKS

Most data traffic still passes over public switched trunk networks (the regular telephone system). In Europe, that still usually means that the national PTT company is the only or main carrier. Although many of the restrictions on the sale of hardware, particularly for data communications, have been removed, the PTT usually controls what equipment may be connected to the network, and has the sole right to set prices.

These networks were designed many years ago with voice communications in mind. Because human speech is perfectly understandable even when limited to a range of 4 kHz, this is the bandwidth of the cable and switching systems used.

Data is modulated onto this essentially analog network by a modem (see Chapter 1 for the terminology) at up to 1200 bits per second (1200 baud). For higher speeds, three techniques are used. Phase changes on the signal are used to give up to 2400 bits per second in both directions simultaneously (usually, slightly inaccurately, called 2400 baud full duplex).

Speeds higher than this are usually achieved by data compression—removing unnecessary bits from the data stream or compressing predictable sequences such as a long string of ones (bytes represented in hexadecimal as FF). The highest speeds are achieved by relying on the fact that the connection is often very much better than the minimum specification. The modem therefore throws data at the other modem at very high speed, but in blocks incorporating additional (redundant) bits. The receiving modem evaluates the check data and accepts or rejects the block. If several blocks are rejected, then the sending modem tries a lower speed or smaller block size.

Once the line reaches the telephone exchange, the signal is usually converted into digital form for long-distance transmission. Some PTTs, for example in Germany, are able to offer higher-grade lines from the local connection (the office where the computer is located) to the exchange. This will be of particular benefit when the rest of the connection is digital (for example, when the PAD is located at the exchange). It is usually worth specifying that a line will be used for data, since most modern exchanges will then treat the signal differently.

Modems are an inherently inefficient way of sending digital data. Modems and analog connections account for most of the problems and errors in any computer network, and EDI systems are no exception. The modem spends much of its call time just establishing the connection—waiting for tones, dialing, waiting for more tones, working out what kind of modem is at the other end, exchanging passwords and so on, before it can send any useful data.

Since much exchange and transmission equipment is shared between various users, the actions of other telephone subscribers or engineers working on their lines can affect one's own telephone service. For the same reasons, the public network is basically insecure: anyone can monitor a call simply by obtaining physical access to any point on the connection.

On the other hand, the public network offers exceptional convenience.

- New connections in Western Europe or North America usually only take a few days.
- Any telephone line can in principle be used (in some countries subject to PTT restrictions).
- It is international.
- Most important, it is a system to which almost everyone has access.

For this last reason alone, PSTN connections will continue to be used widely for data until there is a digital network with equally wide coverage in each country. France is probably the only country that has made a deliberate policy to separate voice and data traffic.

5.3 LEASED LINES

Users who send a lot of data to one other point—another office in the same company, a major customer, or the EDI bureau itself—can have a special line installed between the two. Although still owned by the telephone company, it is leased permanently to the user at a rate dependent on the speed of the line and the distance involved.

A leased line is always available (barring faults). Data can be sent down it at any time without any need to establish a connection. A leased line is in fact a "logical" connection—it is not a physical piece of wire. (In the case of trunk lines, it will in practice be a multiplexed transmission; that is, combined with other signals and then split out again.) However, the user is unaware of this, and can use the link as though it were a physical connection.

Different speed connections are available, up to the very common 9,600 bits per second and beyond. Most of the main network software, including PC local area network (PC-LAN) software, can treat computers at the other end of a leased line as though they were on the network, although with lower performance and sometimes fewer facilities. Users who have many leased lines can use them as a private network.

Most public or shared EDI services include a private network, so that customers can make a local connection from anywhere in the country. Some of these networks (notably the IBM and GEIS networks) are international in their scope, although larger networks will nearly always use digital connections rather than analog leased lines.

The cost of a leased line does not vary with the amount of data sent, so it is usually cheaper than a dial-up link for connections of up to a few kilometers, and where the quantity of data transmitted would involve more than one or two hours a day during peak telephone charges. The gain in reliability and other factors may make the economic threshold much lower. The quality of a leased line is matched to the speed at which data will be transmitted, so reliability is generally much less of a problem.

Security is another important difference: although a leased line does still use common switching equipment and can be monitored without the user knowing, it is much more difficult for another user to connect to the system, and easier to determine if any attempts have been made to tamper with the connection. Password controls should be retained with a leased line.

Beyond a few kilometers, and particularly at high speeds, leased lines become very expensive in most countries. They are therefore not usually economical except for links within a small community of users, or for connecting to a local node (connection point) on a high-speed network.

5.4 PACKET SWITCHED NETWORKS

An inherently much more efficient way of sending the type of messages involved in EDI is to wrap them in an electronic envelope and send them across a digital network.

A digital network does not have the inefficiency of converting all the digital data into analog and back again; nor does it have to maintain a connection between the two ends of the line when it is not being used.

Digital networks work by taking the stream of data and dividing it into packets. Each packet is given a header and a trailer; these identify the source and destination, and add some check data. The packet is then inserted in the stream of data from the user equipment to the first node. At each node in the network, the header data is read and the packet is sent on to the node that will get it closest to its destination (see Figure 5.1).

There are packet switching networks available in most countries, and these are generally linked to the international packet switching system (IPSS). There are also many private packet switching networks, some of which are used by the major EDI services.

The main international standard for packet switching systems is CCITT recommendation X.25. There is a linked standard, X.28, that defines the packet assembler-disassembler arrangements for accessing a packet switching network from an analog connection. It is very common for small and medium-sized users to connect to a private or public X.25 network through an X.28 PAD accessed through a dial-up line, as shown in Figure 5.2. The central EDI system will probably have a direct X.25 connection.

5.5 ISDN

The next stage in the development of the digital network is the high-speed public network, known as the Integrated Services Digital Network.

ISDN subscribers have a socket on the wall (or more usually a set of them) similar in principle to a telephone socket. This offers a high speed connection (usually 64 kilobits/second) that can be used for voice or data. Any analog signal is converted to digital by the device connected to the ISDN network.

As with the telephone network, any subscriber on the network can call any other subscriber. Provided that they have compatible equipment, they can send data or other signals up to the maximum speed of the link.

In some respects, ISDN is a threat to current EDI services, since it allows trading partners to send data to each other very simply and with a minimum of formality. Many EDI users will seek to become ISDN users in due course anyway.

But this is to forget several of the other functions that an EDI system can perform: these include providing security to network users, buffering messages to suit the operations of each trading partner, splitting messages from one company to many different destinations, logging the traffic on the network and confirming that a message has been collected, and possibly also some format checking or conversion. An EDI system is more than a communications channel, it is a part of the company's operations.

And underlying any EDI operation is the concept of common standards and agreed formats. Ease of connection does not remove the need for these, and many of the topics discussed in this book would be equally valid if the whole ISDN system were treated as the EDI communications system.

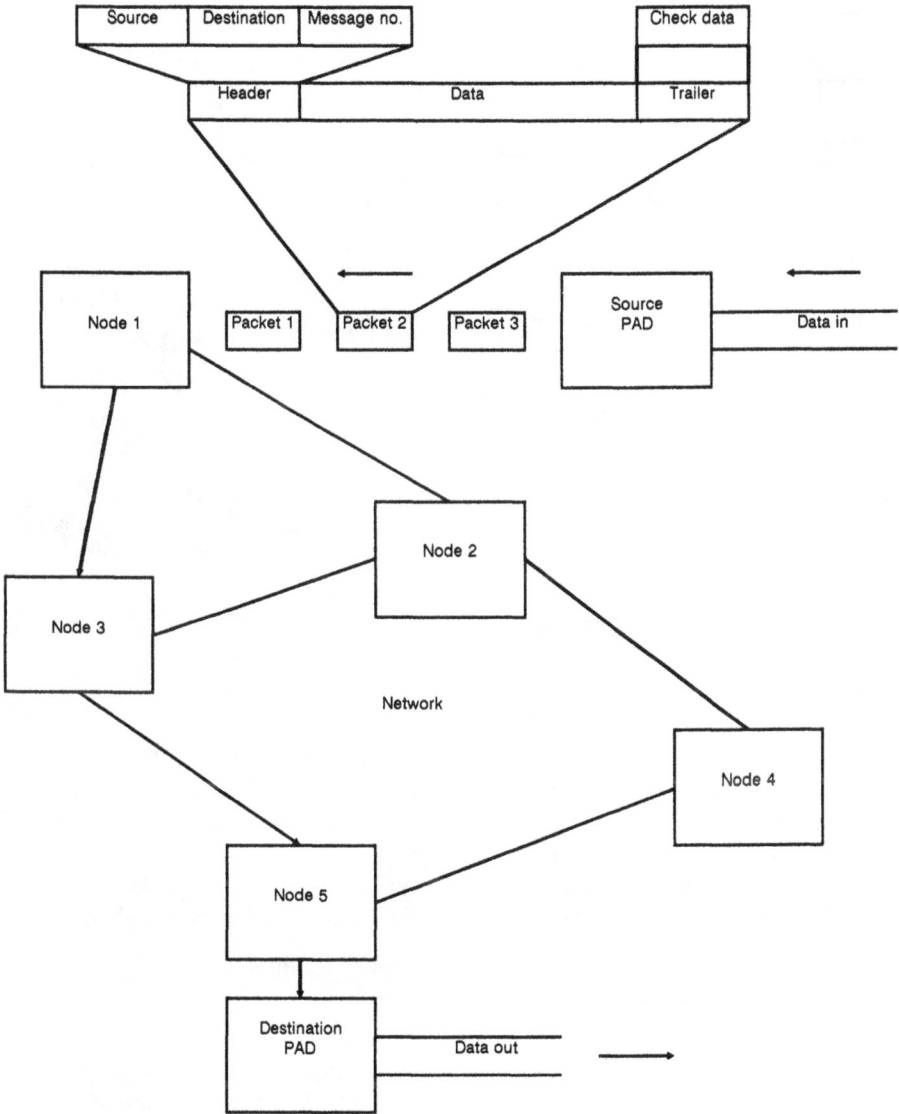

Figure 5.1 Packet switching.

ISDN is today most widely available in France and in Germany, although several other European countries are catching up fast. ISDN is still very expensive today, although it is usually priced competitively compared with other 64 kilobit/second services; however, few users make full use of its data capacity.

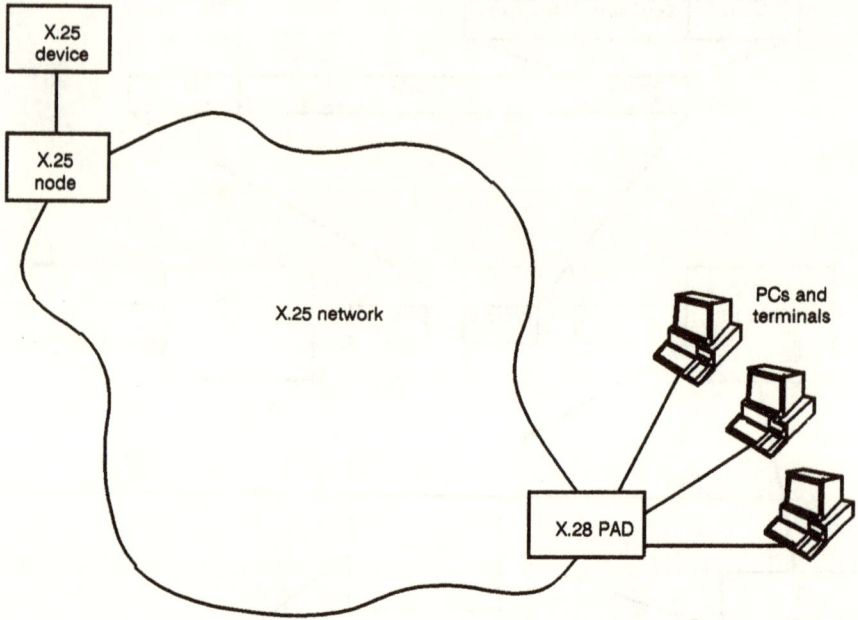

Figure 5.2 X.28 PADs in an X.25 network.

In the longer term, ISDN is likely to have its greatest value for EDI users in what are termed "technical" EDI operations—those applications that fall outside the main commercial services. They often involve passing large files using a pre-agreed EDI structure directly from one user to another.

Another important area will be access to private networks—in order to make use of a high-speed network, users need to be able to access it at the same high speed.

5.6 HYBRID NETWORKS

Few large networks today use only one technology. We have already pointed out that most long-distance telephone connections "go digital" for at least part of their distance.

Whereas copper or aluminium cable is still widely used for the local connection, fiber optics are normally used between exchanges. Microwave links are also widely used, although more in North America than in Europe.

At the beginning of the 1980s, it was thought that satellites would take over most long-distance communications, certainly all international traffic. In practice, the economics of optical fiber technology have improved faster than those of satellites, and so within Europe this is the main form of international communication for voice and data.

Satellites are most widely used today for television, navigation, and other applications where the higher data rates or the special range of a satellite are required. The original aim of the World Administrative Radio Conference, the body governing satellite standardization, was to have certain satellites devoted to television, using one set of standards, and others devoted to telecommunications, using another. This has not worked.

Satellite projects have a very long lead time, so the number of communications satellites in geostationary orbit aimed at Europe will continue to increase throughout the 1990s, even though there is considerable surplus capacity today. The new satellites will have a higher power, and can therefore be used with smaller dishes (20 to 45 cm). One of the biggest drawbacks today is the large dish size often needed for good quality data reception. The increase in capacity may well drive prices down and make satellite communication attractive again for many data communications applications.

Most international networks will use all of these technologies, as shown in Figure 5.3, although the aim is normally to make this transparent to the end user.

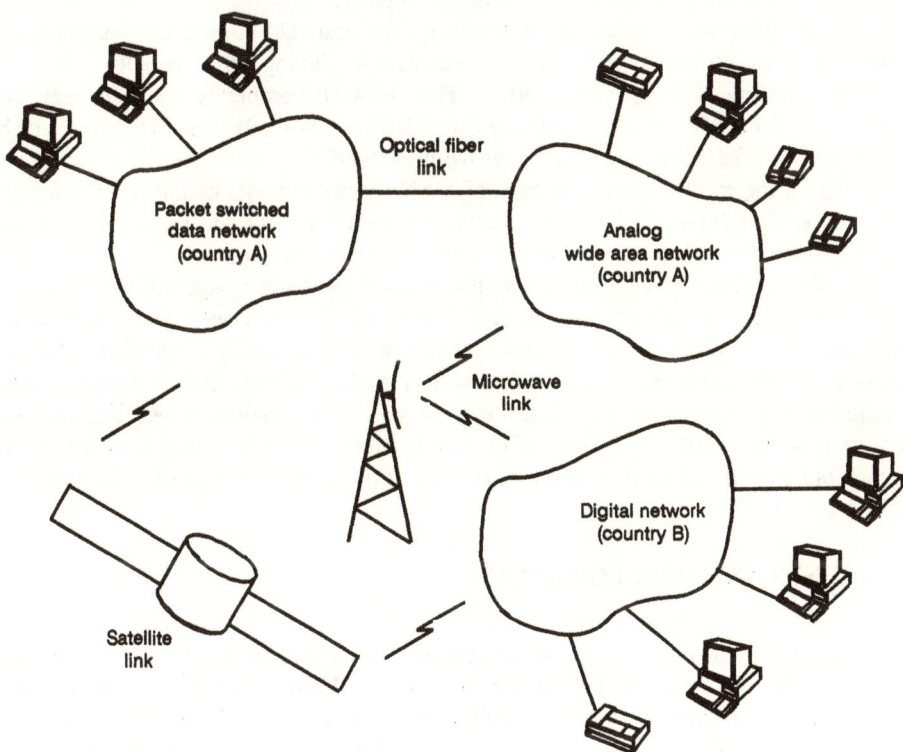

Figure 5.3 Hybrid network.

5.7 VALUE-ADDED NETWORKS

With the exception of the national PTTs themselves, few companies offering network services to the public only transmit data from A to B. Most of them also offer other services: database storage, for example, or switching between different types of network (a gateway). As a minimum, they will offer some added security, through the use of redundant paths, traffic logging, and possibly encryption.

EDI services are one example of such a value-added network, also sometimes called a value-added data service (VADS).

A VAN operator will normally have a computer at each node of his network, often relatively powerful. Each of these nodes will be able to perform or contribute to most of the network functions, possibly with the exception of data storage, which may be concentrated in one or two locations.

Users will be connected, either directly or by dial-up links, to a node of this network, and are then able to gain access not only to other subscribers connected to the network, but also to the services provided by the host computer.

Some networks, such as Swiss Radio's Data-Star, U.S.-based Compuserve, and General Electric Information Services, specialize in linking other people's networks together, creating a "super network." This presents problems of standards and addressing—precisely the areas addressed by EDI, so it is likely that EDI users will be well placed to take advantage of these growing networks.

One of the most important groups of VAN service consists of the X.400 message handling services (MHSs) offered by most of the national PTTs and major data carriers. These are now being linked together to form an international network.

X.400 is primarily designed for the transmission of text messages directly from one user to another; a special envelope, called PEDI or X.435, has been designed to allow EDI messages, which have a different syntax and must normally be buffered, to pass across an X.400 network. You could view X.435 like the customs declaration labels required to send international parcels through the mail without being opened. This standard is not yet widely used, but it does offer an opportunity for EDI systems to access the large population of X.400 users and to make use of the software written for this standard.

5.8 TRENDS AND DEVELOPMENTS

One important trend in the data communications industry is the blurring of distinctions between local area networks and wide area networks. From a functional point of view, the difference is becoming very small, although no wide area network can yet achieve the data rates (several megabits/second) associated with high-performance LANs. EDI users on LANs will, however, be able to take full advantage of all the facilities of a wide area VAN, while EDI techniques may be used more and more within a single LAN.

Companies where different departments make use of different EDI services may have a single gateway computer (as in Figure 5.4), performing the logical and physical conversion between all the different standards involved. Or the LAN technique may be used to allow, for example, the EDIFACT format conversion routine running on a departmental PC to send some data to an X.25 system connected to a central computer, and other data to a leased line connected to a second PC.

In North America, there is support for a move towards so-called metropolitan area networks (MANs): very high-speed wired networks covering an area of several square kilometers. Although there are pilot metropolitan area networks in Europe, the pattern of trading appears to be different, and customers are much more interested in trying to make the public X.25 and ISDN services more widely available and cheaper. The X.435 extension referred to above will probably make the ''X.25 with X.400'' a popular form of communication.

The focus of most data communications network developments in the next few years will, however, remain the value-added network. There is a move towards interconnecting

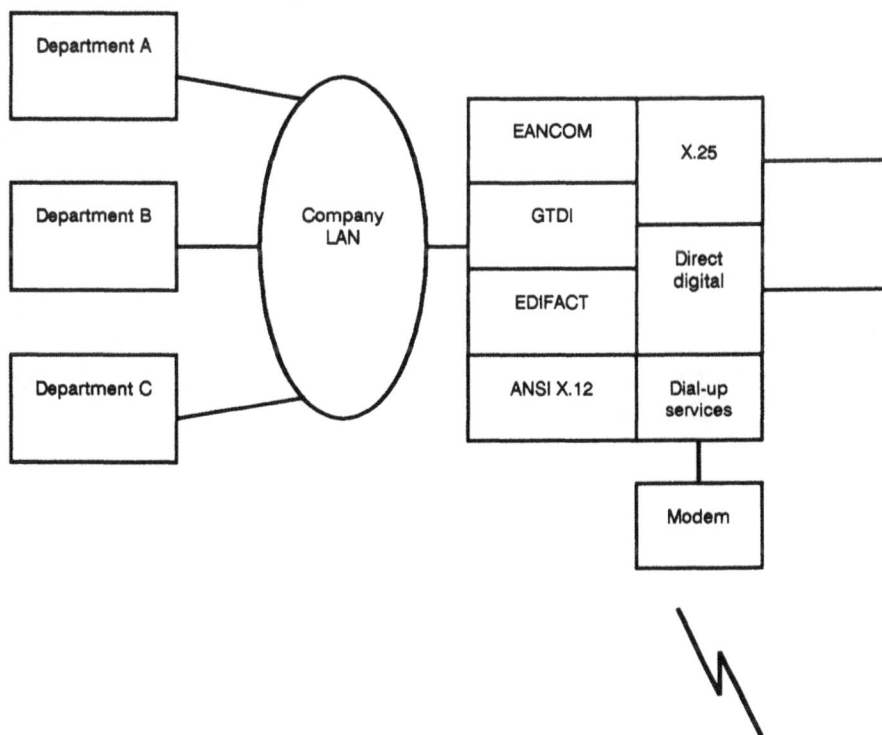

Figure 5.4 Company gateway.

networks that provide a common service, such as information databases. This could well happen to EDI services, particularly as the EDIFACT standard takes hold in many more sectors. Users may then choose to subscribe to one EDI network, and still be able to send messages to users on other networks.

Chapter 6
Central and Remote Hardware

6.1 HARDWARE INDEPENDENCE

One of the key features of EDI is that it is, or should be, independent of the hardware on which it is run. As we will see in Chapter 7, all EDI standards relate to messages, that is the content of the data being exchanged between the trading partners.

Most systems also make use of common open systems interconnection (OSI) standards, which help to define the way the data is exchanged, down to the levels of the electrical signals. None of these standards is specific to any manufacturer or type of computer hardware, although as we will see they may restrict the type of communications hardware used.

We have seen in Chapter 4 that EDI software normally works with what are called "flat" files—simple strings of data divided into records and fields. To allow maximum freedom to exchange data with any other systems, the most common trade EDI standards even restrict the character set used to the set that was used by Telex (upper case letters only, and very limited punctuation).

Most EDI systems can therefore be accessed with any kind of hardware. Indeed, many EDI users have a variety of hardware, or possibly even a network. There are, however, some minimum requirements and some desirable features. It is almost impossible to say what kind of hardware is most suitable for EDI in general, but this chapter will try to indicate some of the areas that should be considered.

6.2 EDI HOST SYSTEMS

Most users are not concerned with the hardware at the EDI host system. But it is actually not very difficult to set up an EDI host, since several companies provide software that can be used in either "host" or "remote server" mode.

An EDI host system is likely to consist of two or more computers for security and load-sharing reasons. These will usually be mid-range or "superminicomputers." Larger systems will usually have more machines rather than more powerful units, unless they also provide other value-added features. Computing power is rarely a limiting factor for an EDI system.

Good communications facilities are essential, and specialized communications processors (known as front-end processors) are often used to make sure that as many users as possible can be serviced simultaneously. When loads are heavy, it is usually best to slow all users down equally but smoothly, rather than simply cutting users out or making performance "jerky." Although a "jerky" performance (obtained by changing priorities dynamically) actually gives a higher throughput, the smooth degradation is preferred by users.

The host system should have at least two connections to the network. If the main connection is a direct X.25 connection, then a second X.25 connection in a different place would be the preferred route.

An EDI host must under no circumstances lose messages. All messages must therefore be stored at least twice, and it is generally best if two different forms of storage are used, to avoid "common mode" or systematic errors. A common combination would be to have main storage on disk, together with a continuous tape archive of all transactions. Alternatively, two sets of disk storage may be used if they are in different locations.

The host will probably also have media conversion facilities (for example, floppy disk, cartridge tape, and half-inch tape), a line printer for the system log, and a number of terminals for operator control.

6.3 USER HARDWARE

EDI users cover the whole range of computing, from the most basic PC user to international corporations with powerful computing networks of their own.

A PC is perfectly adequate for most companies' EDI needs. In many cases, this will be the same PC on which the other applications (such as inventory control or invoicing) are running. In these cases, the only additional hardware required will be a modem.

The next step up is where the company already has a PC network. In this case, the EDI application may run on one of these, or there can be an EDI server, as in Figure 6.1. A server in a network is a common resource available to applications running on other machines in the network. The EDI server may also run other communications tasks, or may be dedicated to the EDI application.

Even for companies with multi-user computers or larger networks, it will often be convenient to use a PC as an EDI server, since it is then detached from the housekeeping operations of the central multi-user system. Direct connection from a multi-user mainframe system into an EDI network raises potential security problems, but may be justified for very large EDI users with direct connections to secure networks.

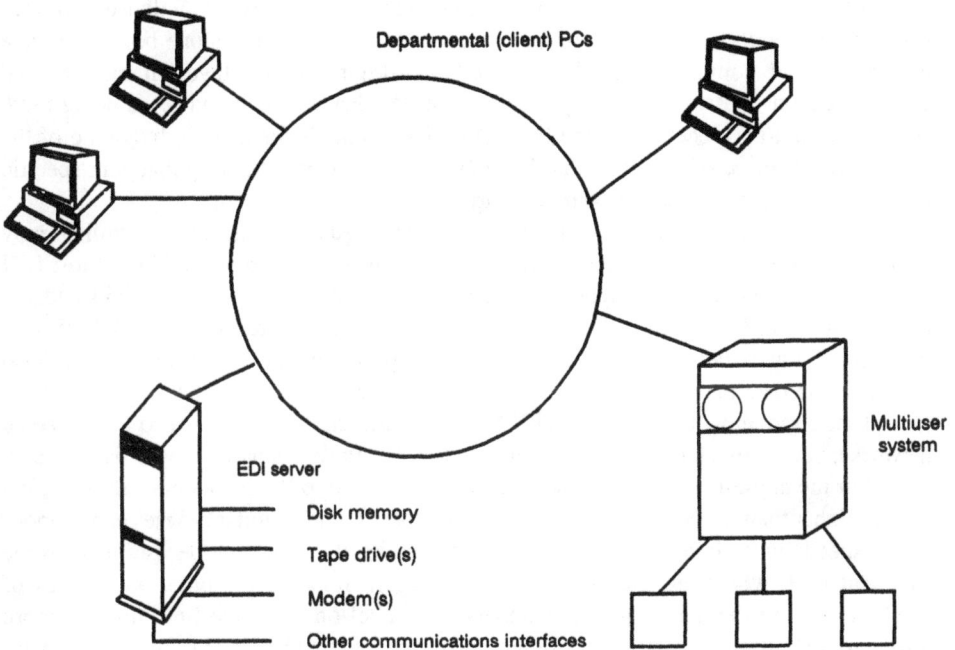

Figure 6.1 EDI server.

As we saw in Figure 4.1, the EDI server will often need to have connections to several different types of network as well as providing different format conversion routines. As with the EDI host system, it may also be necessary for the server to have various types of disk and tape drive.

Unlike other PCs in a typical company network, the EDI server will normally run 24 hours a day. It should therefore be on a maintained electrical supply, possibly with an uninterruptible power supply (UPS) so that loss of power does not cause any loss of data.

6.4 COMMUNICATIONS HARDWARE

Unlike the choice of computer hardware, users have little choice with regard to the type of communications hardware they use once they have selected a network or EDI service. For users starting to use EDI for a trading application (sending purchase orders, invoices, customs documents and the like), speed of communications is unlikely to be a major factor. The ability to connect to a particular system or network will determine the choice of modem or other device.

The simplest systems, running over the public telephone network, will use a conventional modem, either in the form of a PC card or an external stand-alone box. There is a common set of commands for dial-up modems, originated by Hayes in the US, and communications software generally assumes that this Hayes command set is being used. Almost all dial-up modems use the basic ATDT commands for dialing, but some of the commands are more specific to individual modems. It is worth checking that your specific modem type is recognized by your software.

As we saw in the previous chapter, the highest speed the public network will directly support is 2400 bits/second in each direction (the standard known as V22.bis). Most EDI systems will, as a minimum, support error correction to either the V42 or MNP5 standards (V42 is a CCITT recommendation, whereas the slightly more advanced MNP5 is a proprietary, but widely used, standard originally promoted by Microcom). These also compress the data and so have the effect of increasing the effective data transfer rate.

Faster modems, using the CCITT V32 standard, can now be used to access some networks, but are much more common in private networks and smaller closed user groups.

For the highest speeds, EDI users must look forward to ISDN modems. ISDN offers speeds higher than the communications ports on most PCs and similar devices can support. These speeds will almost never be required for trade EDI, but may be useful for some technical EDI, where the file sizes are much larger. Some of the other advantages of ISDN, including the ability to run channels in parallel on the same link, may be more useful to EDI users, although the range of devices available for this purpose is very limited.

Users of packet switching (X.25) networks may also access them through an X.28 PAD using standard dial-up modems, as we saw in Figure 5.2. Larger users may make direct PAD connections.

Value-added networks may offer users direct connections in their premises using a high-grade leased line interface or multiplexer provided by the VAN operator. This solution has a higher initial cost, but because the user is committed to that network for a certain period, the network provider will usually absorb the installation cost, and normally handles most communications problems. Users should recognize that, although this type of connection has a higher reliability than, for example, a dial-up line, there are fewer alternatives if the line should fail.

6.5 SECURITY

As we saw earlier, security in an EDI system can only be achieved by a combination of measures, some relating to system design and others to software, hardware, and procedures.

The hardware features that will enhance security are, first of all, those that help the system's integrity and reliability: uninterruptible power supplies, and backup and duplicate storage facilities. On PC systems, it is now common to have automatic backup facilities, typically using cartridge tape, that require no operator intervention.

Hardware access control features include special devices (usually called dongles) that must be in place before certain programs can be run, or before a terminal or PC can be used. The most powerful—and also the most dangerous—of these store and do not allow access to the boot sector of the hard disk until presented with a password or other key. Dongles should not be used in isolation, but in combination with some software security or password control.

Smart cards are being used increasingly often to provide security features in connection with PCs and networks. Smart cards, which are usually the same size and shape as a bank card, incorporate a processor and some memory. Part of the memory is protected, in the interests of security, and this is usually used for storage of passwords and other codes that can be checked directly by the card. Several designs also incorporate software for decoding DES, RSA, and other encryption algorithms. These features make the smart card a very useful tool in the design of secure EDI systems.

France has been a leader in promoting smart card systems, and it is in this country that these devices are most widely used.

Smart cards are also used in encrypting modems; in this case, the two modems facilitate an exchange between the cards that sets up the keys. Encryption and decryption are carried out by the modems, since the cards are not currently fast enough to perform these tasks at the speed of a real-time communications channel.

Other security hardware includes or has the effect of physical locks, even if they are electronic or software controlled.

Chapter 7
EDI Standards

7.1 THE NEED FOR STANDARDS

When a product is first introduced to the market, its developers usually try to keep all the parts of the design to themselves. They may even protect them using copyright or patents to prevent others from developing competing products.

Even when other designs have entered the marketplace, cooperation on standards is rare. In the early stages of product development, it is often easy to gain a competitive edge by improving performance or by making something easier to use. Competitive standards often appear, and it is in the interests of the market that they should appear.

Sooner or later, the market starts to include a wider range of users rather than just the keen ''early adopters'' who will pioneer any product. This is probably the stage at which it is important for a standard to become accepted, and this stage will come sooner for a ''conceptual'' product than for something that you can feel or touch.

This need for standards from a certain point in the lifetime of a product is one of the characteristics of the product life cycle (see Figure 7.1).

Even in the software arena, designers nowadays recognize, at least in principle, that programs designed to meet published or de facto standards have in most cases an inherent competitive advantage. It is actually surprising how many proprietary EDI products exist (or EDI systems that use no recognizable standard at all), when the published standards are used across a very wide area.

For a ''product'' such as EDI to move into the rapid growth stage of the product life cycle, it is almost essential to have available a wide range of competing software products conforming to a single standard or set of standards. This situation has only been reached in a very limited number of sectors and application areas.

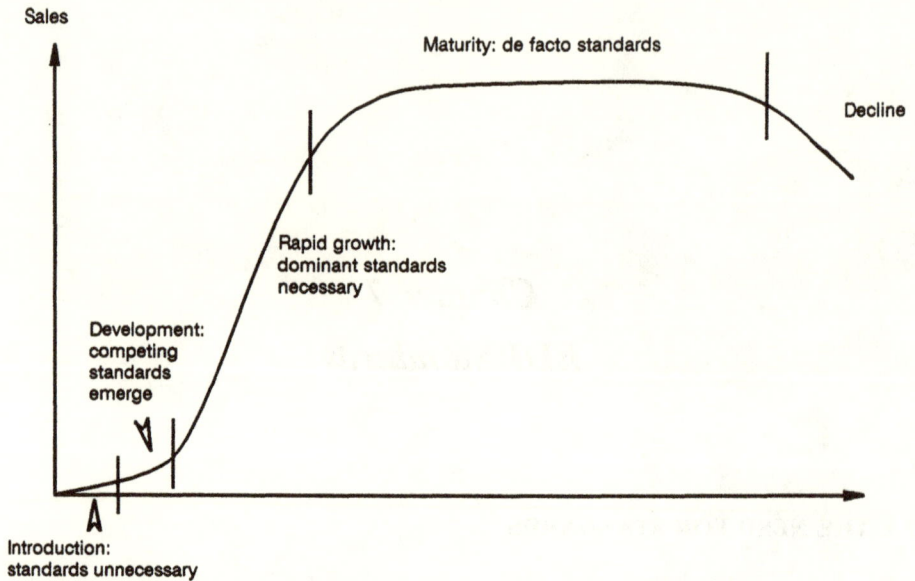

Figure 7.1 Standards in the product life cycle.

7.2 REQUIREMENTS FOR A STANDARD

The special characteristics of EDI impose certain requirements that must be met by any good standard. EDI is still very much a growing area, and so any standards defined now must be general enough to cover future uses, many of which simply cannot be foreseen. More specific areas, such as coding systems, must allow for easy enhancements and additions.

EDI files or interchanges are normally "posted" with no facility for immediate acknowledgement. This means that if the receiving system cannot interpret the message unambiguously or at all, a major problem may arise. An EDI protocol should therefore include a checking mechanism, so that messages can be checked before being sent and, in the case of a central EDI system, on receipt by the central system.

The protocol should also if possible be readable by humans as well as by machine. It is best if both the meaning and the structure are clear when the message is displayed or printed. This is likely to lead to better system design as well as making it easier to provide backup systems and to recover from errors.

7.3 SCOPE OF EDI STANDARDS

A complete definition of an EDI system should include all the system components:

- Minimum hardware;
- Software requirements;

- Communications system;
- Syntax and protocols.

Chapters 4, 5, and 6 cover the first three areas. We now need to look at the standards covering the content of the data that will be transmitted.

As we will see in the later chapters, EDI standards need to cover a very wide range of industry sectors and applications. The most highly developed are those that cover standard commercial transactions: buying and selling goods, transporting them from point A to point B, paying for them, and so on. This is the preserve of the UN/EDIFACT protocols, which we will discuss in much more detail later.

Financial and banking applications originally used proprietary protocols for their EDI systems. They have come closer to the EDI commercial standards as they have become more relevant to bankers and their clients.

Other ''special interest groups'' within the EDI community, such as the oil and shipping industries, have felt the need to extend the simple commercial transactions to cover special messages relevant to their businesses. Nowadays they usually find that this can be done by adapting or extending the EDIFACT message set. They do this by proposing new draft messages to the EDIFACT board.

For technical applications (such as sending computer-aided design output from one company to another, sending marked up documents between lawyers, or sending publishable documents to printers and publishers), wider-ranging standards are being developed.

7.4 UN/EDIFACT

7.4.1 History

UN/EDIFACT has its origins in the early 1980s, when two separate groups were pursuing EDI in Europe and in North America. The European group was a working party of the United Nations Economic Commission for Europe (UN/ECE), while in North America a committee of the American National Standards Institute was working on the ANSI X.12 set of standards.

These two groups were both working to produce a more general standard from work carried out by earlier committees in the 1970s: in North America, the Transportation Data Coordinating Committee (TDCC), and in Europe the TEDI/TDI standards originated by UN/ECE.

In 1985, a decision was made to merge the two streams of work, and the UN/EDIFACT syntax became a full international standard—ISO 9735—in 1987. EDIFACT stands for Electronic Data Interchange For Administration, Commerce, and Transport. The TDCC and TDI standards are still in use, often alongside the EDIFACT set. American users still most often use the X.12 nomenclature.

7.4.2 Structure

The EDIFACT standard covers the way in which the data is organized inside an EDI interchange. The EDIFACT committee also provides a mechanism for standardizing specific messages. There is a separate standard—ISO 7372—that defines the elements of the standard messages. EDI partners can agree on other messages where necessary, but these would be outside the standard. Figure 7.2 shows the family tree of these related standards.

Since the principles used in the EDIFACT structure are common to almost all commercial EDI, it is worth going into them in some detail.

EDIFACT is not in itself complicated. It does use rather a lot of abbreviations and jargon that make it seem impenetrable to the beginner ("Does your UNH include a Common Access Reference?"), but the simple analogy of a set of paper documents, as shown in Figure 7.3, should make it clearer.

Think of an interchange as an envelope into which we are going to put all the documents we want to send today to our trading partner. Into this envelope we may put a number of orders, some signed delivery notes (to acknowledge that we have received goods) and a payment to cover some previous invoices. EDIFACT prefers that we paperclip together all the orders, delivery notes and payment advices respectively before putting them into the envelope.

The interchange header, called UNB, is the front of the envelope. It carries our address as well as our trading partner's, and a date and time stamp. We should also put a reference code on the envelope, and we may put various other references or a password, as agreed with our partner. In the very top left-hand corner we also say what types of forms or codes we have used in the envelope.

The interchange trailer, UNZ, is the back of envelope. It tells the recipient how many forms there should have been in the envelope.

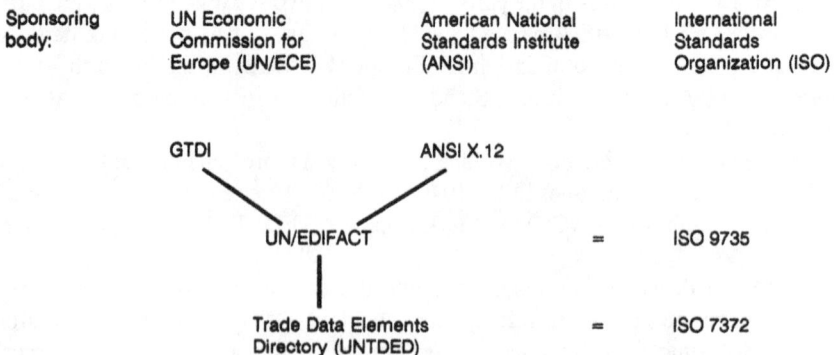

Sponsoring body:	UN Economic Commission for Europe (UN/ECE)	American National Standards Institute (ANSI)	International Standards Organization (ISO)

GTDI   ANSI X.12

UN/EDIFACT   =   ISO 9735

Trade Data Elements   =   ISO 7372
Directory (UNTDED)

Figure 7.2 UN/EDIFACT and related standards.

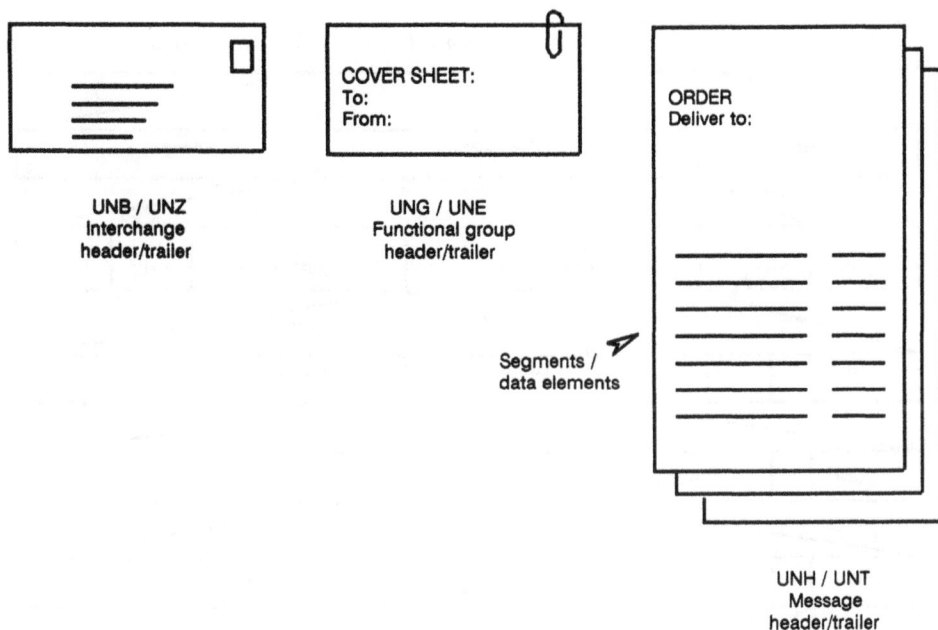

Figure 7.3 Paper analogy of an EDIFACT interchange.

A functional group header, UNG, is the bit of paper at the front of each paperclipped set of forms, which says what type of forms are behind it, and gives some information that is common to all of them. We do not need to use UNGs if we do not want to—we can just put all the individual forms directly into the envelope—but this would mean repeating all the information on each form.

There is a functional group trailer, UNE, that says how many forms there should have been in this paperclip. The message header, UNH, identifies the actual form. This usually carries a reference, known as the common access reference, so that we can add data to this form or make changes at a later date. If we are going to do this, then the form must also carry a revision number that is increased by one every time we change it. A message finishes with a message trailer, UNT.

If we want to depart from the standard characters used to separate the parts of the interchange, then we must specify which characters we are using in front of the whole interchange. This is called the service string advice, UNA.

The structure of messages within an interchange is shown in Figure 7.4.

Messages (forms) consist of a number of segments. Segments are like the boxes on a form; some of them must be completed (mandatory), whereas others may not be needed in every case (conditional). Because the form is electronic, it has more flexibility than a paper form. Segments that are not needed may usually be omitted, whereas other segments

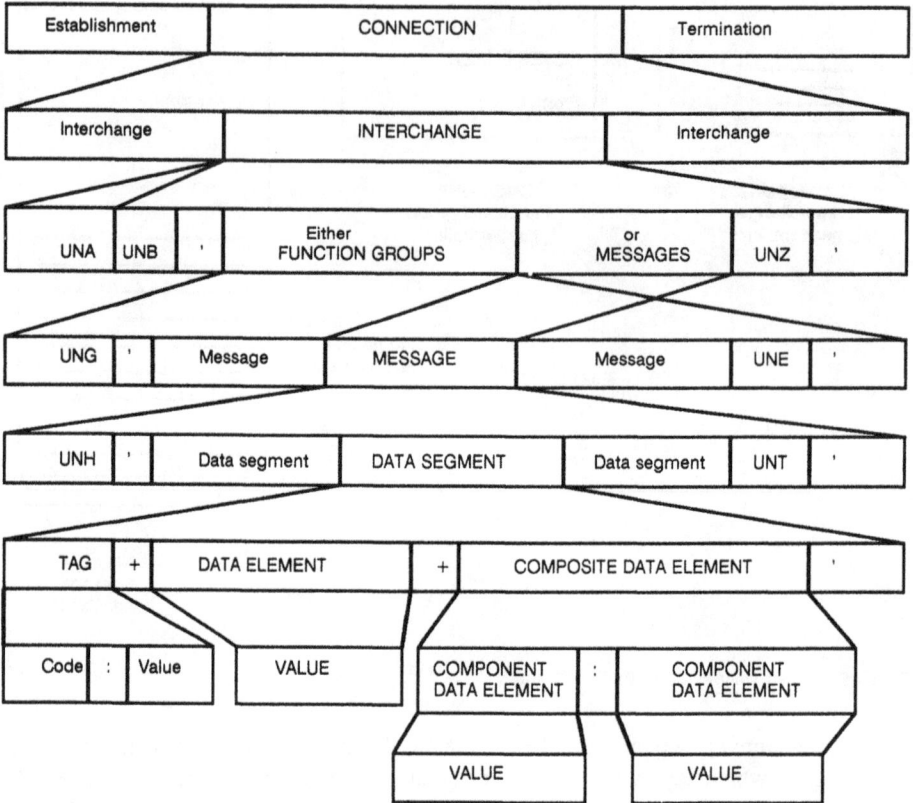

Establishment		CONNECTION		Termination	

Interchange		INTERCHANGE		Interchange	

UNA	UNB	'	Either FUNCTION GROUPS	or MESSAGES	UNZ	'

UNG	'	Message	MESSAGE	Message	UNE	'

UNH	'	Data segment	DATA SEGMENT	Data segment	UNT	'

TAG	+	DATA ELEMENT	+	COMPOSITE DATA ELEMENT	'

Code	:	Value	VALUE	COMPONENT DATA ELEMENT	:	COMPONENT DATA ELEMENT

VALUE	VALUE

Figure 7.4 EDIFACT interchange structure.

(such as the invoice numbers on the payment advice) may be repeated. Groups of segments (such as the item lines on the order form) may also be repeated (for example, one box of pencils, two typewriter ribbons).

The actual sequence of segments, their names, and maximum sizes, are all defined by the message designer. Messages are introduced to the directory as level 0 (draft) messages, then after one year progress to level 1 (draft standard) and after a further year to level 2 (standard). There are currently 18 level 2 messages in the UN Directory, but more are added every time the directory is revised. EDIFACT messages are given six-letter names (e.g., REMADV, PUREXT) that are mnemonics rather than strict acronyms or abbreviations (e.g., COMDIS for "commercial disputes").

EDI users can specify their own messages within their interchange agreements. EDIFACT also helps in this process. There is a standard way of laying out a message design, so that it can be read and understood by others involved in EDI, and interpreted by the various standard interchange software packages. This is illustrated at Figure 7.5.

7.4.3 EDIFACT Directories

As we mentioned earlier, EDIFACT itself provides a structure within which messages can be constructed. The messages themselves, and the content of the data within those messages, are defined in a set of directories. To design or use an EDI system to EDIFACT standards, you have to become familiar with the relevant sections of these directories. They are normally supplied on disk, for use by table-driven software.

The first directory, UNEDMD, is a list of UN standard messages (UNSMs). To run a service using EDIFACT, users should use UNSMs wherever possible. Examples of standard messages include: purchase order; commercial invoice; remittance advice; and customs declaration.

Some of these messages appear more complex than necessary, but on closer inspection you will usually find that most of the extra fields are "conditional," that is, you can simply omit them if they are not required.

The UNEDSD and UNEDCD directories define standard segments and composite data elements that can be used in message designs and offer definitions of the segments used in the standard messages.

The last and possibly most important directory is the Data Elements Directory, UNTDED (which is also published as ISO 7372). This includes codes, definitions, and values for thousands of different fields that could be used in constructing an EDI transaction. Examples of these include: number 5846: Insurance cost; number 7020: Article number; or number 6345: Currency code.

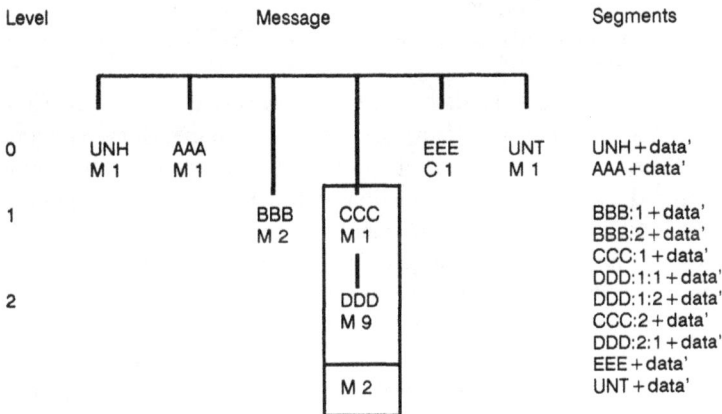

Figure 7.5 Message design example.

It is impossible to give an overview of the scope of these directories, which cover a remarkably high, and growing, proportion of all commercial transaction requirements. Software and message designers have to familiarize themselves with the details, but most users will be able to take advantage of one of the software packages that converts user data into the necessary form.

7.4.4 EDIFACT Developments

EDIFACT is constantly being developed to incorporate new requirements, such as new character sets (accented European, Greek, and Cyrillic), and new modes of operation. The EDIFACT standard itself does not necessarily need to be revised when new messages are proposed or approved, but proposals for new messages are handled by the EDIFACT working groups, and the UNTDED messages and code sets are treated as part of the EDIFACT package.

Proposals from industry groups for new EDIFACT messages or for modifications to the standard are referred to national representatives or to one of the five regional boards within EDIFACT.

Special developments are being considered for interactive EDI (a requirement of the travel industry, for bookings and ticketing), for security and encryption, and for other layers of the open systems interconnection model described in Chapter 4 to be specified within EDIFACT.

7.4.5 Related Standards

EDIFACT uses or refers to several other standards. We have already mentioned the Trade Data Elements Directory. This is an international standard in its own right, approved by the International Standards Organization as ISO 7372 and the European Standards Organizations as EN 27372.

Another very important standard, this time always referred to by its CCITT recommendation number, X.500, is a directory service for use in messaging systems. Although this would be very useful for large EDI systems, it currently has a limitation in that it has been designed to be used in conjunction with the CCITT X.400 message handling system, which is the preferred method of handling transaction messages in an OSI system. As mentioned in Chapter 4, X.400 needs some modifications before it can be used for EDI, and the X.435 standard that makes this possible is not widely used.

As we discussed in Chapter 4, OSI is not in itself a standard, but it does underpin most standards in the data communications area. One further rule that is important to many protocols is known as ASN.1 (Abstract Syntax Notation 1). It refers to the general idea that any information can be encoded as Type + Length + Contents. Although many OSI standards, including X.400, use ASN.1, it is not a general rule in EDI systems; EDIFACT, for example, does not use it.

7.5 TECHNICAL EDI STANDARDS

This phrase may be confusing: it does not always refer to a technical application. But it is the description normally used for all EDI other than in commercial applications.

Whereas EDIFACT is now regarded as the definitive standard for "open" commercial EDI systems, at least in Europe, there is no such agreement over technical standards, which have to cover such widely differing areas as mechanical drawings, electronic circuit diagrams and layouts, parts lists, specifications and legal contracts.

There have been attempts to create such a single standard, but the best that can be realistically achieved at present is a framework within which other recommendations can evolve. Frameworks such as this are usually so vague as to be meaningless or so complicated as to be unusable.

Both of these criticisms have been levelled at the computer-aided acquisition and logistics support (CALS) system of the U.S. Department of Defense, and the similar European Specification 2000M, which both aim to cover all transfers of digital data concerning any product at every stage of its life cycle. As with many large-scale "structural" initiatives, it will be many years before CALS has a significant effect on the normal EDI user or developer. Many individual technical standards are, however, of great importance to users in their respective areas.

7.5.1 Documents

Several office automation suppliers have felt the need for a general representation of a document, common to all word processors, telefax systems, storage systems, and other office products. This allows documents—including titles and revision data, formatting commands, text, and pictures—to be transferred between systems, marked up and revised, and still be printed out or stored on all conventional systems.

The first attempt at this was the Standard Generalized Markup Language (SGML). This is widely used in the publishing sector and allows scope for users to define their own "tags" (formatting commands, for example). But it is basically limited to text.

The Open Document Architecture allows many more aspects of a document to be specified, including its logical and physical structure, revision status, text and graphics, formats and layouts. One powerful feature, very useful for EDI applications, is the ability to refer to standard documents or files: for example, sections of standard text or a company logo. ODA is supported by twelve major office equipment vendors, including Apple, Bull, Digital, ICL, Océ, Siemens Nixdorf, Teles, and Xerox.

7.5.2 Mechanical and Product Design Data

Since the early 1970s, when computer-aided design came into general use, it has been possible to exchange some files between compatible CAD systems. The exchange of data between different systems has remained, however, a largely unattainable goal. The Initial Graphics Exchange Standard (IGES), was specified between about 1979 and 1981, and did make it possible to "translate" basic CAD files (graphics and other product specification data) into a neutral format that could be read by other CAD systems.

In practice, the level of incompatibility between CAD systems is even more fundamental than that: some try to represent objects with three-dimensional images, some as "frames" and some as solids. Their ways of specifying materials and other aspects of a product are equally varied.

The task of creating a standard that would handle all these requirements, without being so complex as to be unusable, has also attracted several groups, among them the U.S. Air Force, French aerospace manufacturers, German car manufacturers, and the European Commission. Each of these groups has produced a standard or enhancement to a standard that attempts to make it easier to exchange mechanical representations and other product data.

Most of these groups have now agreed to work towards a single standard, to be called STEP. STEP, which stands for Standard for the Exchange of Product Model Data, will be the final stage in a series of protocols that aim to cover such areas as conformance testing, geometry and topology, visual representation, and drafting resources, together with "application protocols" for subjects like three-dimensional wireframes and sculptured surfaces.

One of the main aims of a standard like STEP is to cover not only graphical representations of the product (i.e., classic CAD systems) but all aspects of the product throughout its life cycle. This makes it considerably more relevant to intercompany EDI, since many of these aspects will involve not only the manufacturer and purchaser of the product, but other companies as well.

For today, however, EDI of product model data is primarily used by mechanical and structural designers for specifying products and parts. The European aerospace industry and the American defense industry are almost the only users of EDI in this application. Car manufacturers are expected to be the next to follow this route.

7.5.3 Electronic Design Data

Many electronics manufacturers and designers exchange design information with subcontractors and other companies, such as component and mechanical suppliers and printed circuit board (PCB) manufacturers.

A CAD system is so essential to the electronic design process today that many component manufacturers only supply critical design information in electronic form. It is impossible to achieve the level of accuracy and unambiguity necessary by re-entering the information, and so a common electronic format is essential.

The Electronic Design Interchange Format (EDIF) was first proposed in 1983 by a number of major electronics companies, and has been adopted in its present form by the Electronic Industries Association (EIA) and by ANSI. Most leading electronic CAD companies provide interfaces to EDIF, which includes schematic diagrams, netlists (the main input used by most PCB layout programs), PCB definitions, and testing specifications.

EDIF is a versatile standard that provides for a wide variety of library functions and user definitions. A key concept is that of views of a cell—the idea that a cell (the basic design unit) can be viewed in different ways: as a schematic, as a netlist, or as a physical layout, for example.

EDIF relies extensively on English-language words, and has the advantage that both its meaning and its structure are fairly easy to understand from inspecting the file content.

7.6 PROPRIETARY STANDARDS

We will see later that there are many sectors of trade and industry making use of some form of EDI. Several of these, particularly the financial sector, started to exchange computer data within a group according to a common specified protocol before the concept of EDI was generally recognized. Other groups find, even now, that the structures imposed by common EDI standards are limiting, and prefer to use a proprietary standard.

We will refer to various financial EDI systems in Chapter 13. Within the general commercial environment, each national organization involved in the International Article Numbering Association (EAN) promotes an EDI system in its country. In Table 7.1 we list the standards used by each of these countries. It is significant that the French Allegro system, launched as recently as 1989, still uses a proprietary system, while the UK's Tradacoms system permits either TDI or EDIFACT.

Another important area in which a proprietary standard has been used is Organization for Data Exchange by Tele-Transmission in Europe (ODETTE). This is a bar-coding and EDI standard promoted by European vehicle manufacturers. ODETTE now has a policy of moving towards the EDIFACT syntax, although with a number of special messages that still constitute an independent standard.

As EDIFACT becomes more comprehensive and more widespread, it should meet the demands of more of these groups, so that there should be less need for proprietary standards in the future.

7.7 ADAPTIVE EDI

All of the forms of EDI described above could be described as "normative"—that is, they impose a norm or standard to which the user and his data must comply. Although this is a common, probably the normal, approach to computing problems, it is worth reflecting whether this always meets best the users' requirements. EDI should certainly not force people to run their businesses in the way most convenient to the EDI system.

There is an alternative approach, and this is to allow users to define their own communications methods, data structures and message formats. Provided that they can

Table 7.1
National EDI Projects Promoted by EAN Organizations

Country	EDI Standards	Number of Users at End of 1991	Number of Users at End of 1992
Austria	SEDAS	200	300
	EANCOM	–	15
Belgium/	ICOM	73	95
Luxembourg	EANCOM	15	30
Denmark	EANCOM	50	200
Finland	OVT/EDI	800	1,200
	EANCOM	2	Unknown
France	GENCOD (Allegro)	400	800
	EANCOM	15	30
Germany	SEDAS	605	705
	EANCOM	6	20
Iceland	EANCOM	2	20
Ireland	EANCOM	30	100
Italy	EANCOM	30	100
Netherlands	TRANSCOM	500	1,000
	EANCOM	10	150
Norway	"Std records"	2,000	2,050
	EANCOM	21	30
Portugal	EANCOM	–	7
South Africa	SAANA	15	30
	EANCOM	3	5
Spain	AECOM	102	300
Sweden	DAKOM	150	Unknown
	EANCOM	–	100
Switzerland	EANCOM	10	35
United Kingdom	GTDI	>4,000	8,000
	EANCOM	100	300
Total national standards		8,845	14,480
Total EANCOM		305	1,172

Source: EAN. Reproduced with permission.

define these structures and the codes used, the data can be converted by means of a tabular system into the formats needed by their trading partners.

This method works well for simple EDI systems, and has been used for an ordering and transportation data system and for a centralized credit card database system.

Chapter 8
Implementing EDI

8.1 STARTING CONDITIONS

This chapter is written mostly for the user who is considering using EDI for the first time or who has taken one or two steps but is now making a more comprehensive commitment to the technology. Companies in this position are strongly recommended to take a step back and consider the best way to implement EDI rather than immediately taking the most obvious options.

Even those who have been using EDI for some time should, however, go through this type of review from time to time. They may find that they are not getting the best out of the technology. There may have been advances in the technology itself or in the scope of networks or standards available, or other trading partners and organizations with whom they do business may now be better equipped for EDI. New messages and item names are constantly being developed and added to the directories.

We will emphasize throughout this chapter that implementing EDI is not a one-step process. It may require several iterations to produce the initial design. It is then wise to start using it gradually with a number of trading partners who can be relied upon not to complain too loudly if things do not go perfectly from the first day. Furthermore, the scope of the EDI operation and the technology used should be reviewed from time to time.

This must not be seen as a purely technical exercise: EDI will only yield benefits if it is properly integrated into the company's operations. In the case of trade EDI, the operations and finance departments are likely to be the internal "customers" for the service, and, as we will see, they should play a major role in specifying the system.

Because of this need to involve several parts of the organization, EDI does need genuine support from senior management. In the first instance, there will be less support if EDI is simply imposed on a company by, for example, a large customer who insists on sending orders by EDI. In practice, more than half of all organizations make their initial decision to look at EDI because of just such a pressure. But this can be turned to good effect if the advantages to the organization as a whole are recognized.

In Chapter 17, we will again go over the potential advantages and disadvantages of EDI. In Chapter 19, we include a checklist that could be used by anyone charged with producing a justification paper or EDI requirements specification.

Nevertheless, the way a company implements EDI will inevitably depend to some extent on the reasons for considering the problem in the first place. Companies seem to have four main reasons for looking at EDI.

1. Customer pressure. The most common reason is pressure from a large customer or, less often, a dominant supplier. In both the United States and the United Kingdom, large retailers, aerospace and car manufacturers, and government (particularly defense) organizations now take a strong position and insist on virtually all of their suppliers accepting and transmitting certain documents electronically. This is beginning to spread to the larger countries in continental Europe and to other sectors where the use of EDI has become widespread. Users facing this kind of pressure may have little or no option but to adopt EDI. They should nonetheless follow the procedure set out below, in order to avoid problems and reap the benefits, although some decisions will be made for them.

2. Industry moves. The second reason for considering EDI is a concerted effort by an industry or sector association. Industry associations frequently become involved in EDI because they sit on standards committees that are considering connected topics, such as bar-codes or customs procedures. In this case, they may initiate what they see as a way for everyone in the industry to save costs. Or they may be using EDI as a vehicle for sharing costs and giving respectability to an initiative undertaken by one or two dominant companies in the industry. Participating in such a process can make the introduction of EDI much easier, although it is also more open and will probably give fewer competitive advantages. Many of the decisions regarding networks, hardware, and software may be taken by the whole group, although for the company itself the early stages—linking the EDI requirements with the aims of the business—are still critical.

3. Problem solving. EDI may be considered as a solution to a specific problem: a large number of errors in documents, for example, or poor control of the sales administration process. This will again make the economic justification easier, but the process described below should still be followed. There is a real danger in this case that the problem addressed is only a symptom of more general difficulties within the firm. EDI itself is unlikely to solve such problems, although the operational changes that can be introduced around it may be just what is required. It is grossly unfair, however, to expect an EDI project manager to solve all the business' operational or strategic problems.

4. Strategy. The most far-sighted but certainly the least common reason for addressing EDI is that it has emerged as an option or a decision from an overall corporate

strategic review. This is the route that is most likely to lead to a successful implementation. Senior management commitment should be guaranteed, and there should also be a willingness to make changes in other areas resulting from the introduction of EDI. In fact, it may be necessary to reduce management expectations of the benefits to be gained and to temper enthusiasm with practicality.

8.2 THE IMPLEMENTATION SEQUENCE

Figure 8.1 shows the stages an organization should go through in introducing EDI technology. Several of the boxes could be broken down further, and it may often be possible to short-cut some steps or to carry them out in parallel with others. These decisions should be thoughtfully made.

The process described is unlikely to take less than six months. For a larger company or for an EDI operation that will be fundamental to the company, the time period may be much longer.

8.2.1 Terms of Reference

The first stage is to decide the terms of reference of the review and specification process. A key point to consider is the likely scope of the EDI operation: What do we mean by EDI for our company? Which trading partners or other organizations could potentially be involved? What types of communication do we have with the outside world—not just orders and other documents but letters about those orders, telephone calls, instructions to the bank, and so on. Do these originate from a computer? Could they?

At this stage, it is a good idea to keep the potential scope as wide as possible—it is easy to restrict it later. A ''brain-storming'' paper listing every idea, however impractical, may be helpful, although the most-likely candidates for the first stages of implementation should be easily identifiable.

Senior management needs to remind itself of the aims and strategy of the business and to decide whether or to what extent the strategy could be affected by any decisions on EDI. This will help to decide the level and type of person needed in the project team.

The requirements for external communications should be reviewed in light of the strategy. This can be done as a part of the implementation project. It may be necessary to write to major customers and suppliers or to the bank. We have included it here, as it should ideally form part of the terms of reference.

8.2.2 Form Project Team

At this stage the basic goals of the EDI project are set, and we must appoint a team to take the project forward. An external consultant or a person specifically appointed for

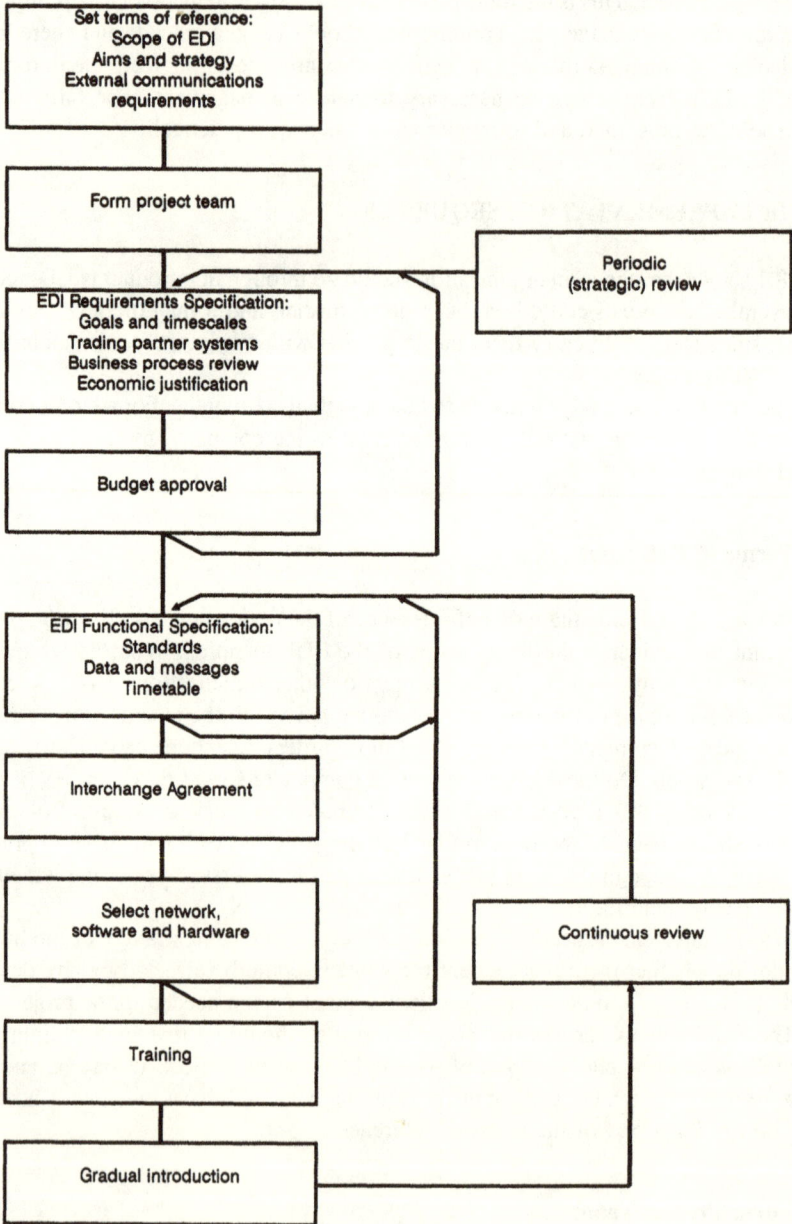

Figure 8.1 Introducing EDI in a business organization.

the job may help here, but the user departments, particularly operations, must be represented and should be active in defining the processes involved. For a larger project, it may be helpful to have an active project group doing most of the work and a review group representing other departments and staff.

One of the team's first tasks will be a general fact-finding mission: reading books and magazine articles, attending exhibitions, probably visiting or calling existing users. Although it is essential for the team to understand the general framework of EDI and the types of companies and applications involved, this is a very subjective process and tends to lean towards the technical side of EDI rather than the operational and service benefits that can be obtained.

8.2.3 EDI Requirements Specification

The project team will start to work toward the EDI requirements specification and probably toward an economic justification document in order to obtain budget approval for the project.

The EDI requirements specification covers much more than just the economic justification for the project. Some will wonder whether it is worth putting this amount of work in before obtaining approval for the capital expenditure involved. The value of including all this information is not only that it helps to determine the opportunity cost—the cost of not pursuing the project—but it also shows that EDI is not just a piece of equipment or software, nor even a management information tool. The requirements specification needs to make the point that EDI is "in line," taking over operational functions. It should therefore be written in operational, and not technical, terms. The requirements specification will refine the scope of the EDI operation, setting immediate, medium-term, and long-term goals. It is important to separate out the key reasons for using EDI, so that these can be given priority in the later stages.

The team must find out what standards are in use by the relevant trading partners, what networks they use, and what others would be available and appropriate. Consideration of other organizations' systems will help to set a budget for the project.

The company's internal systems and procedures must also be reviewed in detail. It is a good idea to draw up a chart of the whole business process—something that many companies never analyze. This often shows repetitive processes or areas where there is no control. But for EDI it is most important to identify where data comes from or is entered into a computer system. Areas to look out for are quotations, orders, job sheets (often known as travelers), and invoices that all relate to the same product or part, and that may be handled separately.

A process diagram should then be drawn up for the company's operations with EDI. At this stage it may be in outline form; it can be refined later. The number of transactions or messages should be estimated.

8.2.4 Budget

By now it should be possible to set a budget for the project. The various cost areas are considered in Chapter 2; they include many personnel and continuous costs as well as the initial hardware and software purchase expenses. Suppliers can be asked for budgetary quotations. It is important to include value-added network operators as well as suppliers of hardware and software, since their prices will interact.

It is always easier to give the costs than the benefits. Savings in manpower are unusual, although often staff can be freed from routine tasks in order to deal with more important work. It may be possible to evaluate the difference in communications costs, which should be to the advantage of EDI. For a comprehensive implementation, the cost of freeing up stock and work in progress can be estimated. The value of earlier payment, where applicable, can also be calculated using the company's cost of capital.

But more difficult is estimating the benefit to be gained from having a more-efficient operation, providing a better service, being more closely in touch with customers, and having the ability to respond more quickly to changes in market conditions. This is why a justification based on a specific strategy is often easier than one purely based on cost savings.

8.2.5 EDI Functional Specification

Once the budget is approved, the team can start to structure the EDI system itself. As well as refining the business process diagram referred to above, we can be more specific about the data and messages to be exchanged, the standards to be used, and the timescale for implementation. These constitute the EDI functional specification; this document will be the basis for most of the work from now on, and will be updated frequently. It must be very specific, and will therefore probably incorporate more technical language than the requirements specification.

At this stage we can start detailed discussions with suppliers and with the trading partners and other organizations with whom we hope to exchange data; this is likely to result in changes to the functional specification. The relative merits of different networks and software systems should be listed and evaluated.

8.2.6 Interchange Agreement

We now start to work on the interchange agreement and the user manual (often an appendix to the interchange agreement). Where EDI is imposed by a customer or supplier, there may be little choice here. Otherwise, we will, if possible, use a single standard agreement drawn up according to UNCID rules (see section 8.3) as a basis, adding only the technical data in the appendices. It is nevertheless advisable to have the agreement reviewed by lawyers in the light of the specific details of each case. The interchange agreement is discussed in more detail later in this chapter.

8.2.7 Select Network, Software, and Hardware

Software, and where necessary hardware, can be selected once a final choice of network (or networks) and standards has been made. A detailed timescale can then be set for the final stages of implementation.

8.2.8 Training

One of the aims of involving users from all departments affected in the project team is to ensure that there is someone in each area who understands the system and is able to champion it. The detailed process of staff training must now begin. It may be necessary to select some staff for more detailed training by a supplier or service provider; however, the general principle is that EDI involves the whole of a given part of the operation, and it is therefore usually necessary to train everyone in that area.

8.2.9 Gradual Introduction

Pilot projects are often undertaken when EDI is first introduced into a sector or application. For most users, it is not necessary to go through a pilot project for evaluation purposes.

It is, however, wise to start by introducing EDI with a limited number of trading partners, and even then to introduce only one message at a time, unless the whole system is very stable. Often both sides can view this phase as a trial, and be reasonably tolerant of each other's implementation problems. The most common problems in the early stages are problems of connection or reliability on the communications link, or parts of the old manual system that have not been sufficiently adapted to the EDI procedures.

Once the system is fully operational with a small number of trading partners, it is usually fairly easy to expand it gradually to others. The EDI system can then be said to be fully "live."

8.2.10 Review

The service should be continuously reviewed in the light of its effectiveness: has it met the aims set out in the original terms of reference and requirements specification? To what extent are manual procedures still required? Have new services or standard messages become available that would be useful to us? Are we still using the most cost-effective communications medium?

As EDI becomes more widespread in the organization, the opportunities to eliminate manual processes and rekeying increase. Bulky reports are rarely acted upon, and can often be reduced to exception reports. The greatest advantage comes when all of a company's suppliers or customers use EDI in all their dealings with the company. This allows a complete layer of manual administration to be removed, although there should always be some fallback mechanism.

The terms of reference and requirements specification themselves should also be reviewed from time to time, perhaps in conjunction with a systems or strategy review. It is important to remember that these are senior management decisions: if EDI has achieved its aims, its profile in the organization will be high enough to command the appropriate level of attention and time.

8.3 INTERCHANGE AGREEMENT

We discussed in Chapter 2 the legal problems associated with using EDI. It is critically important to have an agreed legal framework that specifically covers the electronic exchange of data. EDI partners normally do this by means of an interchange agreement. The interchange agreement has three main functions:

1. It gives the data exchanged a legal status that, in some conditions, it would not otherwise have.
2. It defines what happens and who is responsible if there is any malfunction of the service or if data is lost or destroyed in any way. Several countries have conditions built into their laws that define these for documents sent by post, but for electronic data they have to be agreed between the parties.
3. It defines or refers to the technical standards and message structures that will be used. These are typically incorporated in an appendix or a separate user manual referred to in the agreement.

The bible for authors of interchange agreements is the *Uniform Rules of Conduct for Interchange of Trade Data by Teletransmission* (UNCID) written by the International Chamber of Commerce for UN/ECE. While it was developed for use with EDIFACT, it is equally valid for almost any trade EDI standards.

The rules are very general and cannot be used directly. They state, for example, that an acknowledgement is not required unless the sender specifically requests it and that all parties should keep a log of all messages sent and received for a minimum of three years. Several trade associations and EDI-standards bodies, however, have produced standard interchange agreements based on the UNCID rules that apply to their specific cases. (For example, the Second Edition of the UK EDI Association's *Interchange Agreement* is popular with existing interchange agreement users.) These provide the necessary legal framework but in most cases leave more than adequate room for users to accommodate their own requirements in the appendices or user manual. The user manual will include definitions of:

- The EDI and communications standards to be used, in which versions (and how new versions will be accommodated);
- Which character sets, messages and data elements may be used;
- Conditional fields that are to be treated as mandatory or that are not to be used;
- Network addresses and timings (e.g., messages must be deposited in a mailbox by 2 A.M. in order to be dealt with that night);
- When and what encryption standards are to be used, and if they are to be used, how the keys are to be managed.

The area of encryption and message authentication remains very much within the domain of the individual service and interchange agreement. Although there are very reliable encryption and authentication techniques available, those used by commercial software packages are generally fairly simple. As gateways between different EDI services become commoner, this will start to become an important issue, and one in which we should expect some standardization.

By signing an interchange agreement, the parties implicitly accept that each has software and procedures that are capable of providing the level of control and security required. In the absence of standards on this point, this is an increasingly unsafe assumption.

8.4 PASSING THE BUCK

Once a company has implemented its own EDI system, perhaps under pressure from its customers, the next logical step is to maximize the benefit from its investment by applying pressure on its suppliers to use EDI. This will probably first be in the form of a letter listing the advantages, then perhaps a more strongly worded recommendation. Some companies are able to offer significantly better terms to suppliers who trade with them electronically, while the very largest companies, particularly in sectors where buying power is very concentrated, may decide to deal only with companies who can use EDI. In this way, the advantages of EDI spread through the company, and the technology itself spreads from one sector to the next.

Chapter 9
Applications of EDI

9.1 CATEGORIES OF APPLICATION

We have seen from the previous chapters that EDI is a very flexible tool. The concept covers a very wide range of types of data and companies. Although EDI depends on computer and data communications technologies, it is mostly the way that the technology is used in a particular application that gives it its value.

Every company that has a computer is a potential EDI user. Strictly speaking, EDI applies to structured data: if you use the computer only for word processing, then you could still send letters to your trading partners' computers, but this would be classed as electronic mail rather than EDI. This slightly artificial distinction will probably disappear in due course, particularly as the networks used for electronic mail and for EDI become one and the same through the use of common standards such as X.400.

"True" EDI applications can be divided into three main categories: trade data, technical data, and specialized transactions.

9.1.1 Trade EDI

Trading applications are the aspect of EDI most commonly described, and this is the area that has seen the most spectacular growth in the last decade. Trade EDI is the subject of most of the EDI standards discussed in Chapter 7: GTDI, EANCOM, EDIFACT, and ANSI X.12. It is the main type of data handled by the large networks.

Trade EDI grew along two parallel paths: one led by large retailers and manufacturers who were seeking to reduce their inventories and stock waste as well as reducing paperwork, the other led by a widely perceived need among exporters and transport companies to improve the handling of shipping instructions and contract status. This application was subsequently extended to cover customs documents. These are accordingly the two main sets of messages found in the UNSMs and other standards.

Trade EDI focuses on the standard documentation that accompanies most business purchase and sale transactions: orders, order acknowledgements, delivery notes, invoices, and remittance advice. Each of these represents a message type: in EDIFACT they are known as PURORD, ORDACK, DELADV, INVOIC, and REMADV, respectively.

These messages can then be extended to cover, for example, quotations, payment instructions, customs certificates, notices of disputed deliveries, or invoices (COMDIS); all of which are or can be applicable to most standard commercial transactions. Further messages are then added for more specific applications: documentary credits within the banking sector, for example, or dangerous substances notices in the chemical industry. We will discuss more of these application-specific messages in the chapters that follow.

9.1.2 Technical EDI

As we said earlier, the term "technical EDI" can be slightly confusing. Although it does include the transmission of technical drawings and data, it also covers many other types of file exchange between parties, including documents. The main application is, however, for the exchange of engineering data—the electronic equivalent of drawings, parts lists and materials schedules. Although computer-aided design systems can produce drawings of the objects they represent, they can also describe three-dimensional objects and other characteristics that are more difficult to commit to paper.

Electronics firms are among the main users of technical EDI: they are able to exchange circuit diagrams, printed circuit board layouts, and component lists and specifications, regardless of the CAD system used in each company.

Many applications of technical EDI are one-to-one: they simply represent an exchange of files between two organizations, without the need for any central system or routing software.

9.1.3 Specialized Transactions

Another large category of EDI use—possibly the largest—consists of systems for exchanging one specific type of data, such as payment transactions. These systems are usually run as closed user groups, and very often use proprietary standards optimized for the application in question.

The financial sector makes use of very many such EDI systems, and this is why the banks are considered by some authors to be very advanced EDI users, whereas others see them as lagging in the introduction of EDI. It is true that the banks in most countries have only started to use the common EDI standards such as EDIFACT when pushed to do so by their customers, but their internal use has in fact been substantial for many years.

The whole international payments system depends on a large EDI system covering the whole of Europe and North America and most other technically advanced countries. Credit card clearing is another example. The international card schemes (such as the Visa and Mastercard networks) act as a giant EDI system, clearing transactions in a standard format even though they may have been collected in a variety of formats according to the country where the original purchase was made.

Each of these types of application has its own requirements, and we will consider these briefly before going on to discuss the application of EDI in each industry sector.

9.2 APPLICATION CHARACTERISTICS

9.2.1 Standards

The wider the scope of a system, the greater the need for standards. An organization wanting to offer an EDI system to everyone in a particular industry sector must be able to point to a well-established standard and to a body of software that supports it.

This is even more true if the system is international in its scope, when the body setting the standard should have some international status. Hence the great strength of the United Nations Economic Commission for Europe in promoting the use of the EDIFACT standard.

The drawback with strong standards is that they are expensive and time-consuming to produce, and they can stifle experimentation and fast-moving development. This kind of experimentation is more appropriate for one-to-one applications or small, tightly controlled closed user groups, particularly where everyone is able to use the same software.

With time, such experiments can be incorporated into the standard system. In the UK, a number of pilot EDI systems were set up in the early 1980s, each involving a bank, a manufacturer and another organization. Many of the experiences from these pilot schemes were incorporated in subsequent standards. One of the purposes of the European Commission's TEDIS (trade EDI systems) project is to set up pilots that can be extended into fully standardized systems.

9.2.2 Connectivity

The connectivity of a network or a piece of equipment is the extent to which it can or does connect to other networks or devices—that is, to how many other users or computers it can speak. Network connectivity is critically important to EDI users: there is no point in connecting to an EDI network if it only gives us access to one or two of our trading partners, while another might give us access to most of our current trading partners at home and abroad. Thus, systems that only operate on closed private networks have a disadvantage: although usually faster and more secure, they are more limited in scope.

Trade EDI applications must generally have fairly wide connectivity; many such applications are at least potentially international, and the availability of an international link is therefore necessary. Technical EDI users need only ascertain that the specific organizations with which they want to exchange data are on the same network, while banks and others setting up more specialized systems will normally make use of private networks.

The widest connectivity today comes from the international dialed network, and some users who require a fully international scope must use this. The widest digital connectivity is offered by the X.400 network, and this is why there is considerable interest in the X.435 (PEDI) extensions that allow EDIFACT and similar messages to be transmitted using the X.400 standard. In the very long term, the ISDN network may provide the ultimate solution.

With private networks, one point to look out for is that some systems, while offering wide geographical coverage, are optimized for only one type of hardware or communications protocol. This can cause problems, for example, with PC systems on a 3270 network, or eight-bit systems on Digital Equipment networks optimized for seven-bit operation. Although such problems can nearly always be solved, it is best to be aware of them from the start.

9.2.3 Speed

Users on a "mailbox"-based EDI system can each connect at a speed appropriate to their requirements. For the majority of EDI users, certainly in the early stages, data transmission speed, even at 2,400 bits/second, is not likely to be a limiting factor: most messages are fairly short, and the transmission time might only be a few minutes a day. A high proportion of EDI data transmission takes place outside working hours. Where a dedicated EDI server or similar PC system is being used, no other application is delayed by the EDI system.

Where the system timing is such that data transmission must take place during the day, more attention should be paid to the length of time that transmissions will take. Higher-speed modems (such as V.32) or a digital connection will reduce call charges and will often repay the higher cost of the equipment.

Where the EDI transactions will be transmitted frequently during the day rather than in one or two batches, call set-up time—the time to establish a connection and go through any logon or security procedures—is more important than the actual transmission speed. This is where a leased line or permanent digital connection may pay for itself.

9.2.4 Security

Every EDI network must have a level of security appropriate to the nature of the data it is carrying. Whereas orders for pencils from a school to a stationery company would not be regarded as critical, an instruction to transfer a million dollars from one account to another clearly demands a very high level of security. In between are many types of transaction that, although not in themselves highly sensitive, their originators would not want seen by a competitor.

Many EDI networks insist on some if not all of the data being sent in clear (i.e., unencrypted) form, so that they can perform checks on data integrity and treat different messages in different ways. In these cases, one password field is often sent in encrypted form, and an acknowledgement or confirmation may be requested so that the originator can be reasonably sure that the message has been received correctly.

Any EDI network should incorporate good password and key management and should have very strong checks against unauthorized attempts to access the system or to read files. Many users setting up small private networks or one-to-one connections have inadequate protection in this area. This may not matter when the network is small or connections infrequent, but security must be improved as the system grows.

Any user who sends unencrypted data over a normal telephone line must be aware that this is the weakest link in his system: any telephone line can be tapped with ease. Even many financial EDI systems are surprisingly insecure in this respect.

9.2.5 Interactive

A small number of EDI systems, mostly in the travel sector, need to operate in an interactive mode; that is, the operator waits for a response to a message sent. This calls for a system of priorities, so that interactive users can take priority over background tasks (but not to the extent of freezing out these tasks). Few EDI systems have this facility today. On the other hand, there are many interactive multi-user systems in the travel business and other sectors that have this type of requirement. As a result, many systems are being adapted to accept EDI standard messages, and a whole set of new EDIFACT messages is being developed for this application.

For some EDI system designs, the addition of a priority scheme may be a logical extension of the existing operating system. For the majority, however, it turns the design on its head, since they have been designed around the principle of posting messages to a mailbox. It is likely that most interactive EDI systems will for the next few years be developed from existing interactive systems, or will be completely new designs.

Chapter 10
Retail and Wholesale Distribution

10.1 RETAILING IN THE 1990s

The large volume retailers (supermarkets and other large-area retail outlets) have been among the keenest users of EDI in Europe; they and their suppliers represent a very high proportion of all EDI users. This is particularly true in the United Kingdom, where the buying power of the five largest supermarket chains is so strong that they can easily impose systems on the majority of their suppliers.

Since the 1960s, there has been a steady move towards larger retail units (less than 1,000 square meters is now a small shop) and towards larger chains of shops. Not only can the bigger groups afford the introduction of technology, but they also need such technology for their control purposes and in order to make the necessary savings in administration costs. This concentration of retailing started in North America, but has now affected all European countries, starting with the United Kingdom and eventually reaching staunchly traditional retailing customers in Italy and Spain.

The growth of EDI in retailing is closely linked to the growth of bar-coding. Bar-codes and the "scanning" registers that read them have caused the most dramatic changes in retailing in the 1980s. We discuss this growth and the techniques involved later in this chapter.

Another critical factor is that EDI not only links supplier and customer electronically, but also sends the relevant instructions to transport companies and warehouses, as well as payment orders to the bank. This integration of the whole purchasing process is driven by data that can be collected in real time at the registers: as customers make their purchases, systems in the back office can analyze and predict when the next shipment will be required, which styles are selling fastest, and whether the whole scale of the next months' forward orders should be increased or decreased.

The other big change in retailing since the 1960s has been the increase in variety demanded by customers, particularly in fresh foods and in fashion. This is partly because the most advanced retailers are able to provide an amazing level of variety, and partly because communications and the mass media have shown people what is available. Another important factor is the fact that consumers have developed a desire to differentiate themselves: variety is becoming a human need as basic as friendship or security.

Holding stock has become unacceptably expensive: retailers like to have as much of their stock as possible on the shelves. Stock ties up expensive capital, requires expensive storage space and, worst of all, rots or goes out of fashion. Retailers therefore aim to have their suppliers deliver goods more or less as they will be required; in some cases, this can mean making two deliveries a day to a single shop.

Pulling all these elements together, supply chain management has become fashionable among larger retailers. This idea was often practiced by specialist retailers before; it involves a close contact between supplier and customer, often on an exclusive basis, whereby each keeps the other informed about lead times, expected orders, and schedules. While this still cannot completely prevent "stock-outs," it does ensure that each knows what it can reasonably expect of the other, and there is a shared responsibility for ensuring that the goods are there when required.

Most of this can actually be achieved without EDI, and many small retailers practice all these techniques using the telephone, fax, or other online technologies. The overhead becomes impossible, however, as the volume of trade rises, particularly with a large number of separate suppliers. It is no surprise, then, that trading between retailers and their suppliers (consumer goods manufacturers) has the highest rate of electronic trading in the general marketplace (although, as we will see below, communications between banks are even more intensively electronic).

As a general rule, retailing is still a national business: in the few cases where retailers do cross national boundaries, they usually run their foreign outlets as separate businesses. There seem to be few compelling rationalizations for this: there is probably more in common between the city-dwellers of Frankfurt, London, and Copenhagen than there is between such urban citizens and the farmers of Bavaria, Yorkshire, or Jutland. It is likely that the historical reason lies in the different supply, distribution, and payment systems used in each country.

With the increasing use of EDI on an international scale and as the single European market takes effect, these differences could disappear. In Chapter 18, we discuss further some of the potential implications of this change for both the retailing and EDI businesses.

10.2 BAR-CODING

Data to be transmitted or exchanged by EDI must be structured, using a set of rules and codes agreed upon between trading partners or by all the participants in a single EDI system. For consumer goods, the codes for different products are provided by the barcoding system. EDI in retailing would be possible without bar-coding, but there are so many links between the two that they are in practice always tied together.

Bar-coding is a simple means of automatic identification: a label that can be read by a machine. Although most familiar on consumer goods, bar-codes are also widely used in industrial control, security systems (personal badges), transport systems (way-bills and baggage tags), and in many other sectors. There are many different types of bar-code: some of them can encode letters as well as numbers, while others are very compact, for use on jewelry or electronic components. In retail distribution, however, two main types of code are important: the interleaved two-out-of-five (ITF) format used on outer cases and pallets during the distribution process and the International Article Numbering (EAN) system. EAN is used in over 50 countries around the world, and corresponds with the Uniform Product Code (UPC) system in North America.

The EAN code normally consists of 13 digits (a shortened form of 8 digits is allowed for small packages). Two digits represent the country that issued the code; there are then five digits for the manufacturer, who allocates the next five to his products and registers them. The last digit is a check digit that is calculated from the other twelve and ensures that the reading device has read the code correctly. A humanly readable version is printed below the bar-code. EAN codes are therefore unique to a particular product and size of package and can be used freely anywhere in the system. Retailers may also allocate codes (beginning with a 0) to "private label" packs, but these cannot be used freely in interchange.

A very high proportion of dry grocery goods now carries an EAN code in most European countries. Fresh goods pose a greater problem, particularly where these are self-service or weighed at the time of purchase; these must often be identified manually at the register. Figure 10.1 shows the remarkable growth of scanning stores in Europe from 1985 to 1991.

In other retail sectors, penetration varies not only from sector to sector but also from country to country. Whereas in the United Kingdom, Germany, and Scandinavia the rate of bar-coding on clothing is high, it is lower in France, Italy and Spain. Do-it-yourself (home handyman) goods are almost 100% bar-coded in the UK and U.S., but much less elsewhere. Books nearly always carry an extended EAN code that identifies the ISBN number, but only larger bookshops generally make use of this. Pharmaceutical goods carry their own special code.

Bar-codes can be read by various different devices, either by moving the code past the reader (or vice versa) or by a reader that has a moving beam (usually created by a rotating or oscillating mirror). The most common forms of reader are:

- Pen-shaped wands, which are very cheap but less reliable than other forms;
- Hand-held "guns," which used to contain moving-beam lasers but now are more likely to use a semiconductor array (a charge-coupled device (CCD)); and
- The high-speed laser scanners normally built into supermarket checkouts.

This last form of scanner, which uses mirrors rotating about several axes, achieves very high levels of speed and accuracy.

Number of
scanning stores

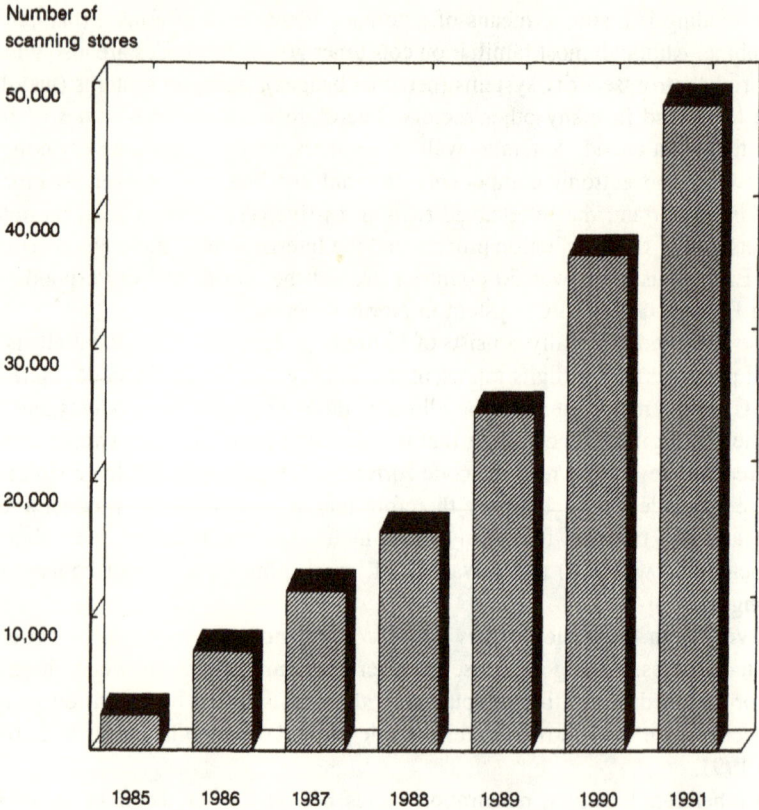

Figure 10.1 Growth of scanning in Europe, 1985 to 1991. Source: EAN. Reproduced with permission.

The use of bar-codes allows a great many further activities to take place, and makes others much easier or more accurate. The important aspect of the EAN code is that it uniquely identifies the product at all stages in the chain, while encoding it as a bar-code means that it can be read by scanners. Information concerning the product can therefore be exchanged between computers within the retail chain, between retailer and supplier, and between each of these and the transport company and warehouse that must also handle the goods.

Additional information such as consignment or batch numbers or delivery addresses can also be represented by bar-codes specified within the EAN system. An in-store computer can take the information read, for example, by a scanner in the goods inwards department and, by reading the bar-codes on the shipping label, can immediately call up the consignment details, including the original order and transport instructions.

The use of bar-coding largely voids the need for retail staff to write down or key in item numbers or descriptions, thus saving time and reducing the number of errors. This is another benefit of EDI.

Scanning bar-codes at the checkout means that retailers, particularly in the fast-moving consumer goods (FMCG) sectors, can keep track from hour to hour of the goods that are selling fastest. This may help to warn of the need to fill up particular shelves, to bring deliveries forward, or to delay them. Such warnings would, however, be virtually useless without the ability to send the revised schedules to the suppliers and transport companies by EDI.

10.3 INTRODUCTION OF EDI

In the retail area, it is the national numbering organizations who issue EAN (UPC in North America) codes who have been the main coordinators of EDI activities. Some individual retailers started to use EDI with their largest suppliers as early as the 1960s. Without the common factor of the EAN code, however, the movement was slow to take root, and the major growth in EDI in retail has taken place since 1985.

As we saw in Table 7.1, each national numbering organization has sponsored or promoted an EDI scheme within its own country, and these have been the main motors for the growth of EDI in retailing. It is largely because of the strength of organizations like Gencod in France, the *Centrale für Co-organisation* (CCG) in Germany, and the Article Numbering Association (ANA) in the United Kingdom that EDI is so widely used. As mentioned in Chapter 7, the standards used in these systems vary from country to country. We will go into more detail on the standards used later in this chapter.

10.4 SEQUENCE OF EDI IN RETAILING

EDI in retailing can operate across virtually the whole scope of EDI, including the transport, customs and financial message types described in more detail in the next few chapters. We will confine ourselves to the central functions of purchasing, delivery, and sale shown in outline form in Figure 10.2.

Most retailers operate under standing purchasing agreements with their regular suppliers. These set down the agreed prices, payment terms, packaging, and delivery methods. A schedule of deliveries is then agreed, whereby the supplier delivers x cases to this store or warehouse on day 1, y cases on day 2, and so on. The first difference made by EDI is that these deliveries are likely to be more frequent and the schedule will change more often—the purchase order change is one of the most frequently used messages. If the supplier himself is taking advantage of the benefits of EDI, he will be able to respond to these demands.

Once a new purchase order or a purchase order change has been sent from the retailer to the supplier, the supplier will probably respond with a purchase order acknowl-

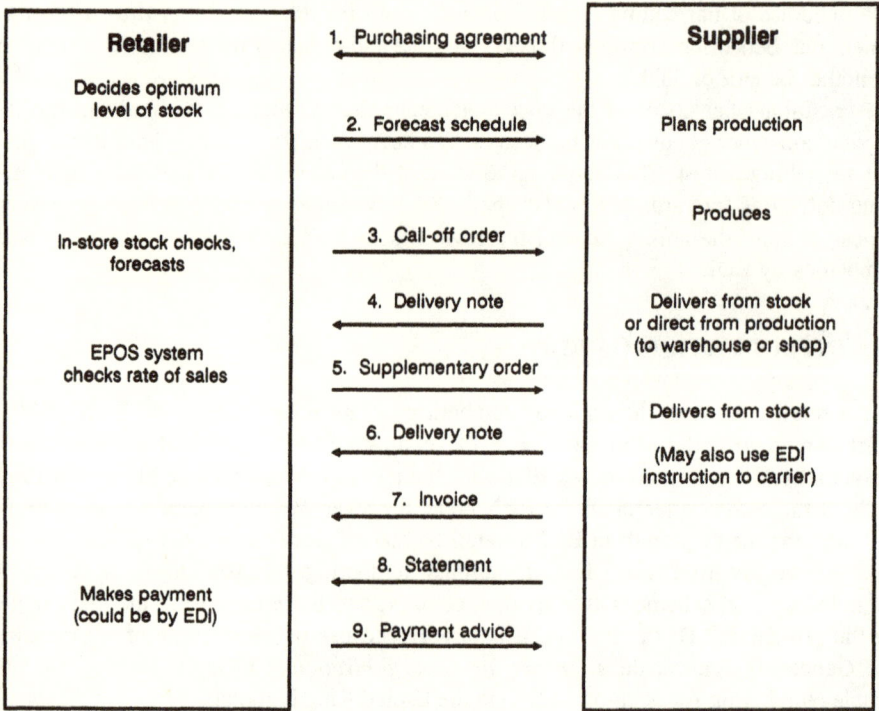

Figure 10.2 Sequence of EDI in retailing.

edgement. This not only confirms that he has received the message, but also that the message has been accepted and entered into his system. Under the interchange agreement, this usually represents a legal commitment.

The supplier now manufactures or procures the goods or simply draws them from its warehouse. He prepares a delivery notice, and if necessary a transport instruction or an uplift instruction, to inform both the retailer and the transport company that the goods are ready for shipment. Where EDI is used to its limits, the scope for delays is very small, and the sequence of instruction and confirmation (order and acknowledgement) provides a useful check that no goods or orders are left behind in the system.

When the goods arrive at the store or warehouse, the bar-codes are read, and the consignment details can be called up on the screen. When the goods have been checked in, a delivery acknowledgement can be generated and sent by EDI to the supplier, who subsequently knows that he can issue an invoice.

Depending on the payment terms, the retailer may now raise a payment order to pay the supplier (we deal with this in Chapter 13).

The consignment details are then used internally by the retailer: they are immediately added to the local stock, and, if the original delivery was made to a warehouse, a similar procedure may be followed when the goods are delivered to the store itself. Records are kept at each stage, so that, even when the outer packaging is broken and individual packs are put onto the shelves and sold to customers, it is possible within limits to retrace the steps they have taken.

The optimal transport arrangements do still vary considerably from one retailer to the next. Many retailers have dispensed with warehouses altogether, and have goods delivered directly to their stores to avoid time delays. Others have all goods delivered to central warehouses; this minimizes the number of deliveries that have to be made and the disruption at store level. A third group uses central delivery for larger, slower-moving items but direct delivery for fresh or locally sourced goods.

10.5 STANDARDS AND MESSAGES

In addition to the EDIFACT and ANSI X.12 standards described in Chapter 7 and elsewhere, there is a set of EDI standards promoted by the International Article Numbering Association, called EANCOM, that is widely used in retail. As with X.12, EANCOM is broadly compatible with EDIFACT but has a slightly shorter procedure for approving new messages and therefore a larger number of messages available than EDIFACT. There is some rivalry between the two standards, although many EDI systems will support both.

EANCOM uses the EDIFACT syntax. The January 1992 release covered 11 standard messages appropriate to the needs of the grocery and other retail sectors:

- Party information;
- Price/sales catalogue;
- Purchase order;
- Purchase order response;
- Purchase order change request;
- Dispatch advice;
- Invoice;
- Remittance advice;
- Sales data report;
- Sales forecast report;
- General message.

Retail EDI got off the ground in the United Kingdom before any of these standards was established. The UK therefore predominantly uses an older standard known as TDI (trade data interchange). Gencod in France and CCG in Germany also use proprietary standards, while, as we saw in Table 7.1, the majority of other national retail EDI systems use either EDIFACT UNSMs or EANCOM.

EDIFACT is gradually being introduced, often alongside the proprietary or national standard. Most of these systems have expressed an intention to accept EDIFACT with, as a minimum, the level-2 (approved) UNSMs. Most of the available EDI software will in practice handle a wide variety of correctly flagged and designated message types.

10.6 WHOLESALING

All of these principles apply to wholesalers every bit as much as to retailers. The goal of providing variety to the end customer with minimal stock held also leads wholesalers to the need for an efficient ordering and distribution system.

Many wholesalers have the same powerful point-of-sale hardware for goods collected by retailers ("cash and carry" operations), and are able to mirror this side of the operation.

Many other wholesalers are more specialized, and will therefore have special requirements. For example, wholesalers who are also importers will have extensive dealings with customs, and may need access to a wider range of EDI networks, or the ability to link suppliers not using EDI into their system. The bonded warehouse requirements of wines and spirits merchants can be very efficiently handled by EDI, so that duty is not paid earlier than is necessary.

Textile wholesalers, for example, may need to be able to send details of designs or other data to their suppliers. Subsequently, their requirements start to include technical EDI, which is also covered in Chapter 14.

10.7 BENEFITS

10.7.1 General EDI Benefits

The benefits of EDI to companies involved in the retail chain are generally the same as were listed in Chapters 1 and 2 for all EDI users. These include:

- Faster transaction turn-around. This leads to better service to customers and, as we will see later, a better use of capital.
- Competitiveness. Speed of reaction to changing conditions is a key factor in many businesses.
- Less paperwork. Large retail businesses are spread throughout their country (less commonly across several countries). Sending paper reports, stock and price lists, orders, and other messages on paper—even if the company uses its own delivery service—is not only time-consuming but also very expensive.
- Savings in staff time. Shelf inventories should still be carried out manually, but these should confirm the stock calculated by the point-of-sale system. The rest of the ordering and invoicing system can in theory operate quite automatically, with each stage triggered off when the goods pass a loading bay or delivery point. In practice, staff check and intervene when needed to modify orders or delivery schedules. In smaller operations, the orders may still be generated manually from manual stock-checks, but the rest of the system can still be automatic.
- Fewer errors. Errors on manually typed documents or misread numbers on handwritten forms were a constant source of problems to retailers. These can be virtually eliminated by a combination of EDI and software controls.

- Reduced payment periods. The retail business has one of the shortest payment cycles of any sector. Cash is still widely used; debit cards, which are increasingly common, give the retailer almost immediate value; and even credit cards have a payment cycle that is shorter than in most businesses. Although many retailers also need cash for investment purposes, this short payment cycle combined with EDI payment gives retailers maximum flexibility as to how to use their cash.

10.7.2 Inventory Savings

One of the key benefits of EDI for retailers is the possibility of reducing inventory. Retailers often have very sophisticated systems for determining minimum stock levels and ordering schedules. These take into account the current stock, average usage for this time of year, variability of usage, and sometimes other variables such as the weather. Among the key factors in these calculations are the risk of the information on the initial stock level being incorrect, the time delay from placing an order to delivery, and the cost of making scheduled or supplementary deliveries.

With EDI, all of these are minimized, so that average stock levels can be reduced. The capital that would otherwise be tied up in stock can be more efficiently used to improve the delivery system and increase the frequency of deliveries, which actually leads to better customer service.

This argument is, however, reversed for the manufacturer or supplier. As we saw in Chapter 2, there should be an overall reduction in the level of stockholding, but some of it is merely transferred from one warehouse to another.

10.7.3 Supplier-Customer Relationships

The supplier gains more from the mutual commitment between retailer and manufacturer that EDI helps to introduce. We say helps to introduce because the commitment is not automatic—the retailer is still free to place orders elsewhere. But EDI does tend to increase the flow of information between customer and supplier, and this in turn makes it easier for each to meet the other's requirements.

The ultimate aim of EDI in retail is to provide the end customer (the consumer) with the variety he or she is seeking, to avoid stock-outs, and to minimize costs. Staff can be better informed, and errors can be avoided.

To give an example, food retailers can gain especially from the short ordering cycle associated with EDI. Fresh foods have a short shelf life—a few days at the most—and have no value at the end of that life. Some of these goods are, in economists' terms, substitutable—if one type of bread is not available, most shoppers will choose another—but others simply represent a sale lost (and a chance that the customer will go elsewhere for that item another time). Some goods need to be really fresh in order to be attractive to the customer. An extreme case is that of mushrooms, which are grown in conditions where they can double in size in an hour just before being picked. In these conditions, a supplier's ability to respond to orders within a very short period is critical to his success in the market. The increase in the variety of fresh fruit and vegetables available in supermarkets in many countries over the last few years could not have been achieved without EDI.

10.7.4 Market Pressure

The benefits described are now so widely accepted in the retail business that the choice for an individual company in the retail or wholesale distribution chain rarely comes down to a fine judgement of the potential costs and savings. Many companies are "strongly advised" by a major customer or supplier that they should use EDI. The issue for them rests wholly on the supplier-customer relationship.

The pressure to adopt EDI can come from either retailer or manufacturer. Much depends on their relative power: in France, the United Kingdom, and Sweden in particular, the largest food retailers are able to exert considerable pressure on the market; smaller retailers are more likely to be pushed by their suppliers. Where EDI is widely used in a retail sector, it can become virtually a necessity for any company wanting to remain competitive in that sector. The pressure in this case comes from the company's competitors.

There are, however, still many retail sectors that are unaffected by EDI, even in the more EDI-intensive countries such as the United Kingdom or Norway.

Chapter 11
Transportation

11.1 THE TRANSPORTATION INDUSTRY

EDI folklore has it that the concept of exchanging data between computers was developed during the airlift of supplies to the American and British sectors of Berlin in the second half of 1945.

The transportation industry (shipping, road, rail, and air) has grown inordinately since that date: world trade has increased a factor of ten in volume and almost a hundred in value, from 300 million tons a year in 1945 to 3.5 billion tons ($2 trillion) in 1990. Trade has grown up to three times as fast as production. At the same time, the cost of transporting goods has fallen dramatically, while the scope and variety of methods available to traders have also increased: it is now possible to ship almost any good anywhere. Today's manufacturers have access to raw materials and suppliers from all over the world, while consumers benefit from a range of products unthinkable fifty years ago.

The oldest and still the most important form of bulk long-distance transport is shipping. Up to the second half of the nineteenth century, nearly all goods transported any distance traveled by water. There were few important towns that were not situated on a coast or navigable river. Speed had always been an important factor in shipping. With the introduction of the steamship, regularity and reliability also came into play.

It was not until many years after the railways were introduced that they became an important part of the transport chain. This took longer in Europe than in the larger land mass of North America. Road transport, which offered more flexibility in destinations and scheduling, entered the picture only in the 1960s, and air transport did not become an important factor until a decade later. In the 1990s, air transport accounts for 10% of the value of goods transported but only 0.28% of its volume. Air transport, as one would expect, is mostly used when time is critical or when the goods are of such high value that the cost of transportation is insignificant. Figure 11.1 shows the relative importance of the various forms of transport used in international trade.

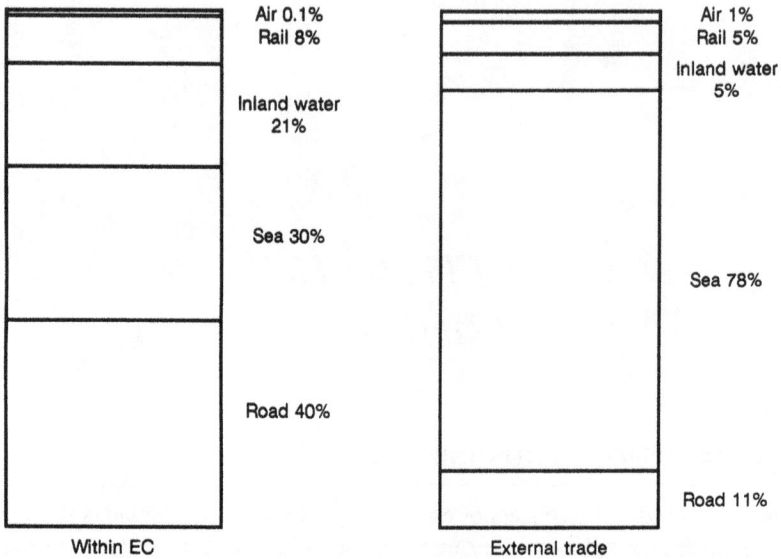

Figure 11.1 Comparison of sea, rail, and air transport in international trade. Source: Eurostat. Reproduced with permission.

11.2 SYSTEMS AND COMPUTERS IN THE TRANSPORTATION INDUSTRY

The simplest form of transportation is the mail, or parcel delivery service. This generally public service is often thought of as old fashioned and people intensive, but there are significant exceptions to this. Most mail sorting is now carried out automatically, and many of the more advanced systems can adapt the delivery services to the needs of the day's traffic.

For domestic trade, rapid door-to-door delivery services are increasingly common. The transportation industry makes the majority of its money not from consolidated loads—made up of consignments from several different companies, going to different addresses—but from direct deliveries on behalf of one company to another.

Each of these modes of operation has its problems: in the case of consolidated loads, the vehicles must be routed carefully to minimize the distance traveled and the delay to the end customer. With direct deliveries, the transport company has the even greater problem of "empty legs"—having to travel in one direction with no load at all.

A major problem faced by all road delivery companies is that of calculating and scheduling driver hours. Routes must be calculated to allow drivers sufficient rest periods and overnight breaks. Computers are increasingly used in the coordination of transport operations in order to minimize all of these problems. Route planning software for road vehicles is now becoming so accurate that transport companies can indicate the expected arrival time of a delivery to their customers within 15 minutes. This means that a suitable crane, fork-lift truck, or team can be standing by, saving precious time waiting for the goods to be off-loaded. Packages or documents are bar-coded and scanned at each collection and delivery point. Vehicles are now frequently fitted with printers so that drivers can receive changes to their routes or print out documents in the cab. Some are fitted with computers that can, for example, revise routes based on actual times and distances.

Communication with the central computer is today usually on a private mobile radio (PMR) frequency. It is likely that after its introduction, digital cellular telephony (also known as *Groupe Spéciale Mobile*, or GSM) will take over the lion's share of this market.

Most consolidated loads, for smaller items at least, are collected at a number of centers, delivered to another center, and then redistributed. The management of these networks of collection and delivery points is a task in which the computer can assist greatly. EDI is ideal for this operation because of its intermittent nature. In practice, most companies have their own proprietary systems for this function, introduced before the relevant standard messages came on to the scene.

Rail freight operates in much the same way as road transport. Today, most goods start and finish their journey by road, with the exception of some very major loads and bulk transport: coal, cement or other chemicals, for example. The systems used by the railways are now very similar to those in road transport, indeed one set of paperwork may cover both modes of transport. In most European countries, rail freight is highly computerized, and EDI in one form or another is therefore the general rule.

The airlines, however, operate in a very different way. In North America, a significant proportion of air freight is carried by specialist freight carriers; these are not a major factor in Europe. The European systems are based on cross-border traffic and there is a high degree of coordination between the systems used by most of the major airlines.

11.2.1 International Trade

EDI is widely used in domestic transportation, particularly in applications such as retail distribution, where it links in with other EDI systems. But it is in international trade that the benefits are most marked.

Goods in international trade are frequently delayed for hours or even days for customs checks at borders. Even when the goods are able to travel unhindered, the cost of paperwork accounts for a high proportion of the expense. Mistakes are common and are often caused or compounded by handwritten documents and language problems. Some consignments arrive without any documentation, which causes major problems.

With EDI, customs clearance can, in principle, be completed in advance, with only a short physical check carried out at the border. Errors are very much reduced, and the

EDI directories themselves act as international dictionaries. Customs authorities in most countries have warmly embraced EDI, and this is one of the fastest growing areas of EDI usage. EDIFACT is very much the strongest standard in these applications.

Ports and airports are themselves very complex systems that pose their own management problems. When EDI reaches a high level of penetration among all shipping companies, this will considerably ease the difficulties of handling the variety of freight that passes through. At present, however, the number of companies able to enter full shipment data by EDI is very low, and so investment in EDI by ports and airport authorities can only be made with an eye to the future. As we will see later in this chapter, many have done so.

11.2.2 The Single European Market

One factor destined to have a major impact on international trading within Europe is the advent of the Single European Market. While many elements required to call the European Community (EC) a true single market are still many years away, trade in most goods was effectively freed from tariffs and restrictions as of January 1, 1993.

There are now few financial implications in trading across EC borders. Domestic value-added tax returns replace import VAT for trading within the Community, and the documentation required for goods crossing these borders has now been reduced to a single form, the Single Administrative Document (SAD). SADs can be computer-generated—this is an ideal application for EDI. The EDIFACT CUSDEC/CUSRES messages correspond to the requirements of SAD.

11.2.3 Progressing Shipments

Given that the physical movement of goods does take time, one of the biggest advantages EDI offers is the ability to check the progress of shipments. This is particularly important where the goods form part of a chain of supply being closely managed by EDI or other means. As a part of a comprehensive logistics management program, EDI is invaluable. Warehousing systems can also feed data to and from an EDI transportation system.

11.2.4 Freight Forwarders

Freight forwarders are the brokers who handle shipments on behalf of trading companies; they act as the interface with the transport companies and, in international trade, with the customs authorities.

The EDISHIP system was set up by forwarding agents and carriers; it enables an agent to ''offer'' consignments to carriers and find out who can offer the fastest or most economical service meeting their requirements for that load. EDISHIP runs on the INS/

GEIS system (see Chapter 15) and is rapidly being extended to cover all the main European countries and carriers.

11.3 EDI PROJECTS IN TRANSPORTATION

Existing EDIFACT standard messages and draft standards meet many of the requirements of exporters, carriers, ports, and customs authorities. In particular, the CUSDEC (customs declaration) message can be used at several points in the shipping process to pass information on the goods' value and country of origin to and between customs authorities.

The CUSRES message is used by customs to give clearance. Both CUSREP and CUSCAR are used by carriers to identify the consignment and mode of transport, while other messages allow users to check the progress of a shipment. The process is shown in very simplified form in Figure 11.2, although there are many variations on this general pattern.

Although several of the initial EDI projects in the transportation industry were set up by carriers and port authorities (e.g., Heathrow Airport's LACES system), EDI has been introduced into the transportation industry on an international scale largely at the instigation of the customs authorities. The Customs Cooperation Council (CCC) has made a firm commitment to introduce EDI in parallel with the streamlining and simplification of customs procedures in Europe. In 1988, the CCC decided to adopt EDIFACT as the standard to be used in international customs messages.

Analyses of the transportation industry's attitudes to EDI give mixed results: while the importance of EDI is not denied by anyone, some feel that many transportation

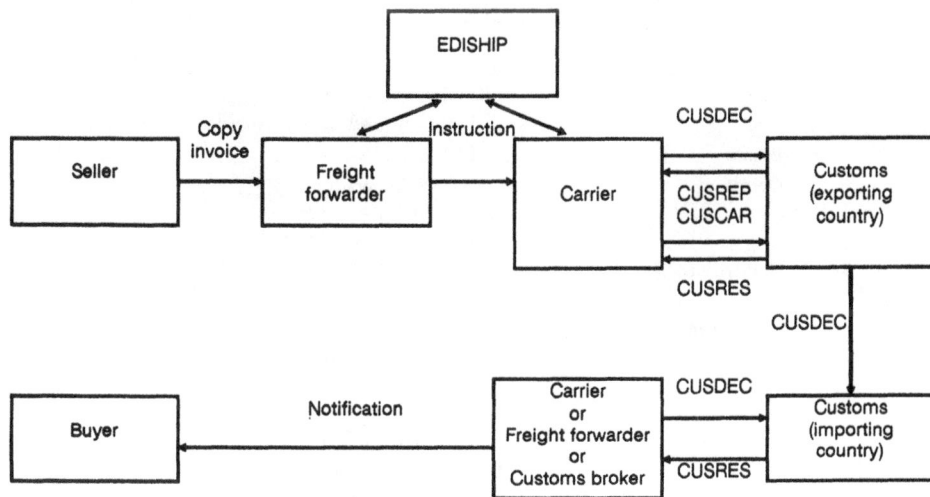

Figure 11.2 Customs EDI messages.

companies remain somewhat unsure how to react. Other surveys and well-informed commentators say that EDI has been embraced by a majority in the industry, and that usage is only a matter of time.

The importance attached to EDI can be seen from the large number of individual projects run by separate port and airport authorities (see Table 11.1), while would-be users' ambivalence is measured by the very patchy use that many of these projects receive. This is at least partly because the whole industry today revolves around the freight forwarders who arrange shipping on behalf of customers. They book loads, complete documents, arrange customs clearances, and chase the progress of shipments. This group, which would be in the forefront of any large-scale move to EDI, feels threatened by the technology, which could make it easier for customers to deal directly with transport companies and customs.

One of the freight forwarders' big advantages at present is that they are onsite in the ports and airports; if documents are exchanged electronically, it does not matter where they originate. The process would become less personal, so that a good relationship with the local customs office would be less important. Their knowledge of the arcane world of customs documents and procedures has already been undermined by the introduction of simpler procedures for the most common routes; it could in the future be encapsulated in the software of an EDI application package.

The support offered by customs authorities makes it certain that EDI will take hold in this area, although probably with some delay. It has been suggested that some authorities will shortly only accept routine customs declarations in electronic form.

Table 11.1
A Sample of Major European Transportation EDI Projects

Country/Area	EDI Project
Belgium	SEAGHA (Antwerp)
	SADBEL (Brussels airport, Antwerp)
France	SOFI 1 and 2 (ports)
Germany	DAKOSY (Hamburg)
	LOSTE/KOMPASS (Bremen)
	ALPHA (Frankfurt airport)
Netherlands	CARGONAUT (Schiphol)
	INTIS (Rotterdam)
United Kingdom	ACP90 (airports)
	SCP85 (Solent ports)
	FCP80 (Felixstowe)
	DISH (shipping)
North Sea	EDDIE/UNICORN (Ferries)
Scandinavia	DEDIST (major ports)
	TDL (Gothenburg)
Europe	DOCIMEL (railways)
World	GALILEO/AMADEUS (airlines)

11.4 BENEFITS OF EDI IN TRANSPORTATION

The form of EDI used in the Berlin airlift would have been a very different operation from what we now know as EDI, but it probably incorporated one key factor: eliminating the need to enter product codes and descriptions repeatedly.

Key entry is a particular problem in the transport industry because the staff doing the entry generally have no knowledge of or contact with the product they are dealing with. Errors are therefore very common: a shipping clerk who enters a value in Italian lire rather than in dollars may cause his customer to pay 600 times too much duty.

Many of the present and future EDI projects in the transportation area are effectively giant databases. The messages are often inquiries to the other party's database for rates, availability, routes, or the progress of consignments. The benefit offered by EDI is access to data.

Paperwork is already being reduced in the international transportation arena. In domestic transport it is mostly computer-generated, so that the move to EDI will simply avoid printing out the document.

The transport industry will benefit internally from the additional available data: route planning is already improved, and the new systems allow operators to fill "empty legs" and to make up full loads. Operational efficiency is generally improved.

For the transportation industry's customers, however, the biggest benefit will be the possibility of removing a complete layer of administration (that previously largely dealt with the paperwork) and dealing directly with transport companies and customs authorities. For the transport industry itself, this will also be an advantage in the long term: it is nearly always better to be in direct contact with customers, and there must in the end be a beneficial effect on rates.

Chapter 12
Manufacturing

12.1 EUROPEAN MANUFACTURING IN THE 1990s

It is constantly repeated in most European countries that manufacturing industries are in decline. In fact, although many heavy industries (including some of the most labor-intensive) are producing less now than they did at the end of the Second World War, total output has grown steadily since 1975, as is shown in Figure 12.1. The shape of industry has, however, changed greatly. The industries that have suffered most have been those that were unable or unwilling to provide customers with the variety or quality of goods that they wanted, or that failed to take advantage of new technologies that would have allowed them to reduce costs. Whereas in the 1940s mechanical engineering accounted for the bulk of all manufacturing in Europe, the trend gradually moved towards electrical engineering, and then to electronics. In the most advanced countries in Europe, all three sectors are now similar in size, with electrical-engineering manufacturing losing ground faster than mechanical-engineering.

Heavy manufacturing generally has a low use of EDI, partly because the business is dominated by small numbers of large orders, and also partly because of the generally slower pace of change in this sector. Other mechanical and electrical manufacturing companies are gradually introducing EDI; many of them are suppliers to the electronics or retail industries, which are both heavy EDI users.

The automobile industry is one of the largest manufacturing sectors, producing 12.1% of output in Europe (see Figure 12.2). Later in this chapter we describe the EDI scheme that covers a large proportion of the European motor industry, although not yet all applications.

The aerospace industry, which encompasses some of all three disciplines, has always held itself apart from the rest of manufacturing. Defense manufacturing has been cut back significantly since the early 1980s and more particularly since the end of the cold war, which was marked by the breaking up of the former Comecon defense alliance.

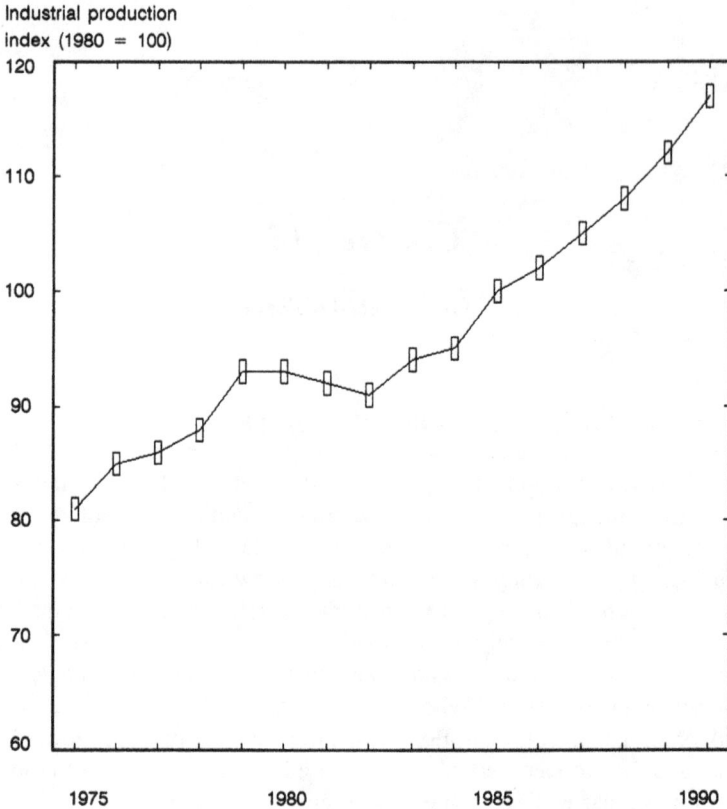

Figure 12.1 Manufacturing output in Europe, 1975 to 1990. Source: Eurostat. Reproduced with permission.

Manufacturers supplying goods to NATO defense ministries are gradually being required to supply full data on all products supplied in a standard format, currently on magnetic tape, to meet a U.S. Defense Department standard called Computer-aided Acquisition and Logistic Support (CALS). The original aim of CALS was to coordinate weapon systems design, procurement, and maintenance by having all suppliers contribute to a common database of knowledge about the product. The CALS principles can also be applied to any large-scale design and manufacture project involving several suppliers. The long-term aim of CALS is to make any form of data available to a service engineer or user in the field through a central database that will be kept up to date by the manufacturer using EDI.

% share of industrial
production in Europe

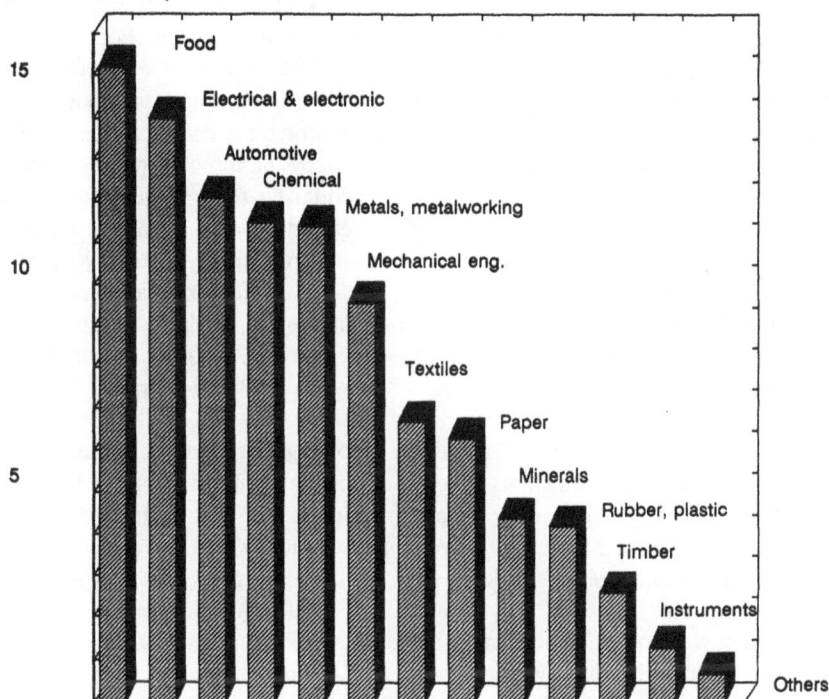

Figure 12.2 Sectors within the manufacturing industry in Europe (1990). Source: Eurostat. Reproduced with permission.

12.2 FACTORS FOR SUCCESS

Ever since J-J Servan-Schreiber discussed "*Le Défi Américain*" (The American Challenge) in the late 1960s, the main factor governing the economics of success in manufacturing has changed from scale to variety. In the 1960s and early 1970s, large-scale manufacturing plants enabled producers to undercut small local manufacturers in almost every speciality. In those years of continuous growth, few markets were truly saturated, and a lower-cost product could nearly always out-sell a higher-cost product.

In the 1990s, with continuous growth no longer the norm, many markets are felt to be saturated, and manufacturers are better able to sell by differentiating their products from those of their competitors. Consumers, greatly affected by the spirit of the communications age, no longer want to be seen as a homogeneous group and demand a higher level of variety from their suppliers. The same spirit affects companies in all lines of business.

Even if the EC has produced a legal framework in which goods may pass freely from one country to another, this has done little to eradicate differences in lifestyle, attitudes, or purchasing habits between consumers in different corners of the Community. The need for variety is increased by the existence of twelve or more sets of national retail groupings, with or without national legislation to reinforce the need for change. Variety at the consumer level translates into smaller batch sizes and more frequent tool changes for the manufacturer. EDI makes it possible to administer these frequent changes, but physical and operational redesigns are also often required in order to make them possible.

The pace of technological progress has increased dramatically. In the 1960s, product lifetimes were expected to be ten to twenty years, and it often took a year or more to bring a product from the initial design stage to the market. Today's key products are considerably more complex, yet they rarely have a sales lifetime of more than five years and must be brought to market in six months if they are not to be overtaken by the next generation.

Manufacturers have to respond to this by becoming more nimble: taking advantage of new technology in their manufacturing processes, changing models more often, and even producing more variants during the production process. This conflicts directly with the 1970s goal of long, stable production runs; but with today's technology it is possible, with careful design, to achieve quick market reaction at a cost lower than experienced with the large-volume plants.

This new wave of manufacturing process design is often associated with Japanese manufacturers, since the three large Japanese car manufacturers and some electronics companies have used them to great effect. We discuss some of the specific techniques later in this chapter.

Two computer aids have contributed greatly to making the new trend towards flexible manufacturing possible. The first is manufacturing resource planning (MRP) II software, which gives production planners the information they need to operate a just-in-time production system. The second is EDI.

12.3 STANDARDS IN THE MANUFACTURING INDUSTRY

Not very long ago, one of the main problems facing manufacturing industries was the lack of coordination between standards bodies world-wide and the absence of international standards. In the past ten or fifteen years, a lot of progress has been made in this direction: not only have organizations like the International Standards Organization become much stronger and more effective, but there are now several new international bodies for coordinating standards work, in particular the European Standardization Center (CEN) and its electrical and electronic counterpart CENELEC.

Although manufacturing standards are still a long way from international agreement, the fundamental step of adopting one Standard International (SI) set of units has been largely completed within Europe.

Other, more detailed standards are less unified. As a general rule, there are now more differences between sectors within manufacturing than there are between manufacturers within one sector in different countries.

Within any one sector and country, there is generally a wide range of standards available for products and processes. Tables of international standards allow users to find equivalents in other countries, while many of these national standards are now cross-referenced to CENELEC and other international sets. These common standards already form the basis of a common language for describing products.

In faster-moving areas such as electronics, however, proprietary standards are still common. One of the least standardized areas is that of data structures, as we saw in Chapter 4. This has a major effect on the ease of introducing EDI for computer-aided design applications. There are many different proprietary standards for CAD and CAM (computer-aided manufacturing) data, and it is only with the help of EDI standards such as the new generation of STEP (described in Chapter 7) that this type of data can readily be exchanged.

One further factor that is starting to have a big effect on the exchange of data between companies is the introduction of comprehensive quality assurance standards (ISO 9000 and its equivalents). ISO 9000 requires a very high standard of record keeping. Where a company's suppliers are not all ISO 9000 registered (i.e., they do not necessarily keep the quantity of data required), the registered company must obtain all the data at the time it purchases the products, and must then store the records itself.

To use ISO 9000 correctly, it is also important to track all batches and to detect any possible sources of error before the faulty products enter the production line. This can be achieved manually, but it is made very much easier by computerized production and stock control, and easier still if this control system is fed with data by suppliers using EDI.

12.4 BAR-CODING

Although bar-codes are much less widely used in manufacturing industry than in retail, the existence of bar-codes and the fact that they are available for use as an easy means of linking a physical product to a more detailed record on a computer database has certainly affected the way modern production control systems are designed.

The actual codes used will vary from sector to sector (see Table 12.1). Consumer goods that will be sold through retail outlets are most likely to carry EAN codes. This all-numeric code was discussed in Chapter 10. For industrial goods, however, most manufacturers have preferred to retain the older alphanumeric part numbers (containing both letters and numbers).

Table 12.1
Examples of Common Bar-Code Standards

Code Name	Data Encoded	Main Application
EAN	13 digits numeric	Retail traded goods
UPC	12 digits numeric	Retail traded goods in North America
ITF	Any number of numeric digits	Traded goods (especially outer packaging)
Code 39	Variable length alphanumeric	Industrial use
Code 128, code 93	Variable length alphanumeric	Longer messages, dot-matrix printers
Codabar	Numeric, special characters	Blood banks

EAN codes cannot contain letters, and the number of codes that can be issued is in any case limited. Manufacturers who do not have the constraint of retail customers normally choose an alphanumeric code, known as Code 39. This is, for example, the code used in the ODETTE system.

Manufacturers of small items such as electronic components and jewelry need very small labels, and they have developed special high-density codes and, in some cases, special readers to read them.

It is often also necessary to identify jigs carrying a part or boxes and pallets stacked with several components. Special bar-codes are often used for these purposes: possibly the same Code 39 symbol as was used for the original part or an ITF label as is common in the distribution industry. These labels are often read by fixed beam readers on a conveyor line that read the labels as the boxes or pallets speed past.

In all these cases, the purpose of reading the bar-code is to link the product with the more detailed information available in the central database. Where goods are being sold from one company to another, it is EDI that transfers the data from the one central computer to the next.

12.5 JUST IN TIME AND RELATED TECHNIQUES

12.5.1 Manufacturing Process Design

Manufacturing process design has been turned on its head in the last twenty years. Whereas in the 1960s it was believed that the key to successful manufacturing was to keep a steady flow on the assembly line, it is now generally found that this leads to poor morale, variable quality, and waste.

Waste comes from many different sources: not only from faulty production (rejects) but also from scrap, excess, or unwanted production. One of the best ways of minimizing waste is to produce only when the product is known to be required. This is just in time (JIT)—one of the key elements of the principle that is now known as lean manufacturing.

Other changes in design philosophy include a recognition that a plant must be designed to produce as many different models or parts as possible, and that tool changes and other delays must be minimized. This is known as flexible manufacturing.

There is also a need to shorten the design cycle, and to improve coordination between suppliers and customers over design details. This implies a regular exchange of design data in the process called simultaneous engineering. All these principles are described in more detail below.

12.5.2 Just in Time

The principle of just-in-time manufacturing is to perform one stage of a manufacturing operation only when it is required by the next stage. Items are only called off the end of the production line as they are sold, and previous stages are scheduled to be completed just in time to allow the next stage to proceed. Figure 12.3 compares a just-in-time manufacturing process with a conventional production schedule, under which goods are produced to a weekly production plan (unless varied by special conditions, such as a rush order or the lack of a component).

In pure JIT, subcontractors and parts suppliers deliver parts directly into the production chain, at the time and place they are required. In this way, the end manufacturer carries no parts stock at all; there is no cost of holding the stock, no administration, nor stores function. In practice, there is a minimum economical delivery size for most items, and it is usually desirable to have a single goods receiving point, where quality assurance checks are also carried out. These buffer stocks should, however, be minimized, and the first items of any batch should proceed immediately to production.

In the case of British car manufacturer Rover, suppliers actually deliver to a distribution center close to the factory that holds a stock of up to five days' supplies. Pallet loads of parts for two hours' production are ferried to the production line. This is a dramatic improvement on the situation ten years ago, when several months' supplies of some parts were held.

Just-in-time production dramatically reduces inventory levels, not only for parts bought in from suppliers, but also work-in-progress inventory: the piles of half-finished goods often left forgotten around the factory floor. The American ''Big Three'' car manufacturers estimate that they reduced their inventory by $1 billion during the 1980s.

But JIT's biggest contribution may often be in reducing waste: if items are only manufactured when there is a demand for them, then many fewer will be surplus to requirements or lost behind a pile of more recent batches.

Far from increasing production lead times, JIT can often decrease them significantly; because the whole manufacturing process is designed around demand, customer requirements can always be fed into the production chain at a suitable point.

Many car manufacturers now gear all their manufacture to end customer orders: it is no longer necessary to wait for a batch of dark blue two-liter models with specific

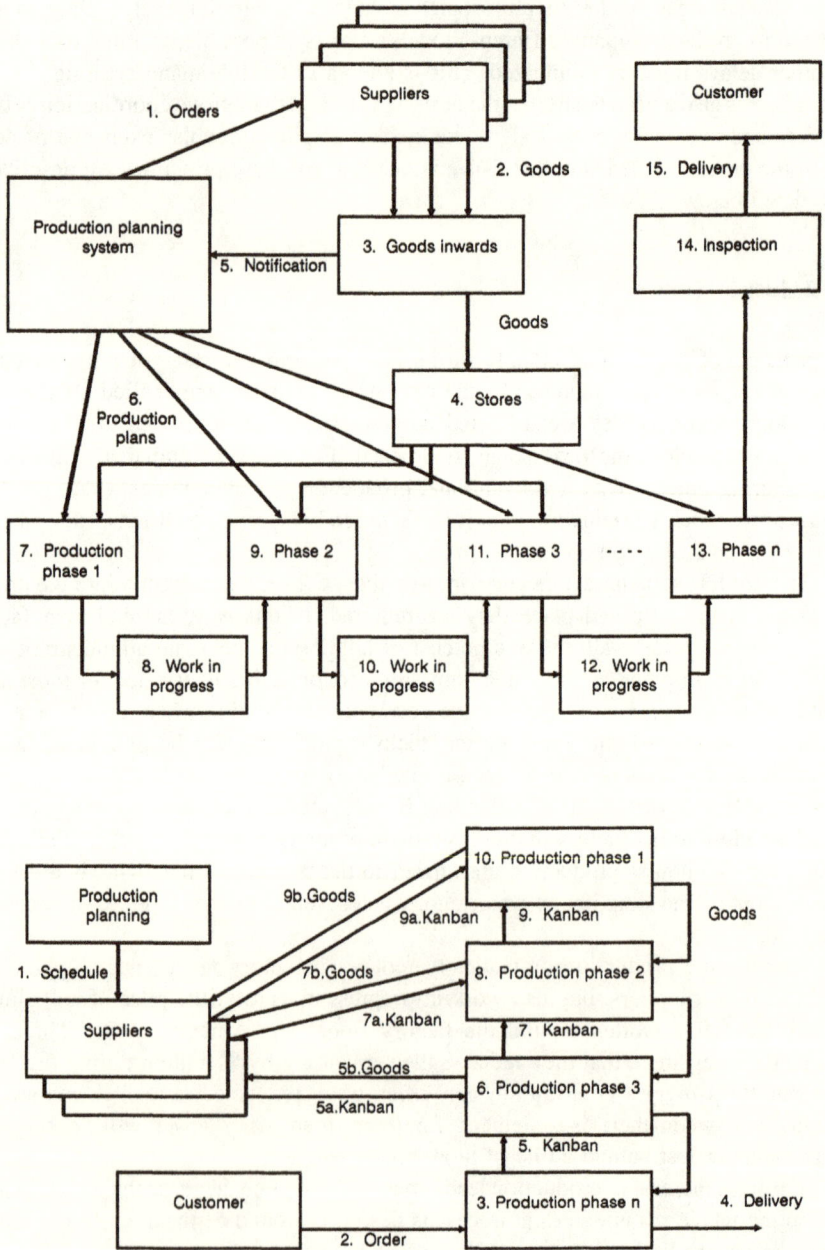

Figure 12.3 Just-in-time manufacturing versus conventional manufacturing.

extras, since the car is simply identified in the manufacturing process the day after it is ordered.

With JIT, ordering is driven by production, not by administration or purchasing. As well as reducing overheads, this cuts down on errors caused by staff not knowing the items they are dealing with.

Just-in-time production is only possible with EDI. Not only must the manufacturer have a well-developed logistics and production management computer system, but he must also be able to keep his suppliers informed of production plans and call in precisely timed deliveries. Parts suppliers are sometimes given delivery times to the nearest 15 minutes.

12.5.3 Kanban

A *kanban* is the flag that triggers the next shipment in a JIT system. The word is derived from the Japanese "signpost" that, in the first implementations of JIT, was placed at the point in a batch when the next batch had to be called up. With EDI, the kanban becomes the electronic message that is passed to the supplier calling for the next delivery. Kanban and JIT must, of course, also be used internally; it is no good if all suppliers can deliver at a quarter of an hour's notice, but internal departments will only supply against a weekly schedule.

12.5.4 Flexible Manufacturing

Less closely linked with EDI, but nonetheless important in the overall scheme of manufacturing process design, is the ability of a plant to change its output quickly. If there is to be no waste from overproduction, there has to be a way of changing production quickly or the manufacturer will suffer from the other great source of loss: people and machines standing idle.

With modern machine tools, molding machines, and filling lines, combined with the ability to download CAM data quickly from a central computer, most individual machines can be adapted to produce different items as quickly as the tool itself can be changed. (For some big presses, this used to take many hours or even days). For a complete plant to be able to manufacture flexibly, however, the changes must be coordinated accurately, so that raw materials or subassemblies are ready when required.

The production control systems for carrying out the corresponding operation with subcontractors that can accomplish this are likely to be well adapted to handling and producing EDI data.

12.5.5 Simultaneous Engineering

Design data is now frequently exchanged between companies in electronic form, often on magnetic tape or floppy disk. With EDI, it is possible for this process to be managed effectively. With the introduction of interactive EDI, it could even take place in real time.

With the increasing need to shorten the time delay inherent in getting new designs to market and the increasing complexity of designs and manufacturing processes, it is critical to manage the exchange of data between manufacturers and subcontractors. If the finished part and the manufacturing process are designed alongside one another, then parts can be designed in such a way that their manufacture can be optimized. This is the process known as simultaneous engineering.

Frequent exchange of CAD-CAM data involves large volumes of data. At analog line speeds, these exchanges can take hours. This form of EDI is likely to remain uncommon until ISDN or other high-speed telecommunications are more widespread.

12.6 THE AUTOMOBILE INDUSTRY

As we saw earlier in this chapter, the automobile industry is one of the largest single sectors within manufacturing industry. It is a highly competitive sector, and the pressure on costs is intense. Because it is also relatively concentrated, the car manufacturers are in a strong position to impose standards on their suppliers.

Within the automation field, the car manufacturers were the first to try to adopt standard interface standards between computers and related equipment. The Manufacturing Automation Protocol has been at the heart of all open systems work since its inception at General Motors in the early 1980s. EDI was the next area introduced by the car manufacturers, in the form of the ODETTE system in Europe and its equivalent, the Automobile Industry Action Group (AIAG) in North America. The vehicle manufacturing industry founded the Organization for Data Exchange by TeleTransmission in Europe in 1984 to set standards and, to some extent, to provide systems for EDI in vehicle manufacture. It is supported by the European Community. There are now over 1,000 manufacturers participating in the scheme in eight countries: Belgium, France, Germany, Italy, the Netherlands, Spain, Sweden and the United Kingdom. There is an organizing committee in each country, usually within the automotive industry trade association.

ODETTE sets standards for bar-coded labels used on vehicles and parts, including a special multiple identification label that defines all the characteristics of a completed car: vehicle number, engine type and number, national standards, color and other parameters. But its long-term (and now its main) role is the establishment of standards for the exchange of messages between car manufacturers and parts suppliers and between different departments within the same manufacturer. At the end of 1992 there were 26 ODETTE EDI messages, covering all the stages of the manufacturing and trading cycle.

The current standard ODETTE transport label is an *A5* sheet incorporating six different bar-codes. These, together with ten other fields on the label, are transmitted between supplier and customer by EDI, so that all the necessary information on a vehicle or part can be obtained by scanning one of the bar-codes and looking up the resulting information on a screen.

Prior to the introduction of ODETTE, the German car manufacturers' association, VDA, had developed its own set of EDI messages; these are still used alongside ODETTE and now EDIFACT messages. Ford, for example, still uses predominantly the VDA message set. ODETTE messages may be used under either the UN/GTDI or EDIFACT syntaxes. Most have up to now used GTDI, but there is a slow migration to EDIFACT. The Swedish car industry, for example, uses EDIFACT rather than GTDI, while most of the German industry is moving directly from VDA to EDIFACT.

Ford and GM both took the step of providing the software for their parts suppliers' end of the EDI link, suitable for PCs or common minicomputers. This helped their major suppliers to adopt EDI, with the effect that the companies were able to insist on this route for supplies of all their critical parts.

12.7 THE ELECTRONICS INDUSTRY

As in many other cases, the electronics industry is the keenest customer for its own products. While other companies nearly all suffer from some technophobia somewhere in the decision-making process, electronics companies sometimes go to the other extreme, using technology where manual systems would be preferable. Within the computer industry at least, companies have access to knowledge about the possibilities of EDI, and implementation may simply be cheaper. Perhaps as important, though, is the fact that this is a relatively new industry, and its managers take it for granted that processes change; they are often visionaries who will seek out changes, and are less bound by existing procedures and customs.

Probably the two largest EDI users in the world are computer companies: IBM and Texas Instruments. IBM has some 2,500 trading partners world-wide (900 in Europe). Its applications include ordering and invoicing, manufacturing scheduling, customs and CAD-CAM. Any IBM supplier can be paid electronically, while any dealer can choose to be invoiced by EDI. IBM's use of EDI includes its own plants; in this case, a special process allows the company to monitor the ''quality'' of the demand pull process by which JIT orders are fulfilled. Supply problems can be resolved and the delivery parameters adjusted accordingly. This process handles 29,000 EDIFACT delivery schedules per week within Europe alone.

Like many manufacturing sectors, electronics is essentially an international business. Products are assembled from components manufactured all over the world. Even in an EDI-intensive sector such as this, the absolute penetration of EDI among smaller suppliers is very low, and because they may all be situated in different countries, they may well use different standards.

The existence of an international standard such as EDIFACT has therefore two advantages: not only does it increase the likelihood of being able to trade electronically with a particular supplier, but it also means that it should be easier to persuade a supplier to adopt EDI, even if he is geographically very remote.

The EDIFICE and EDICUG special interest groups within EDIFACT are active in promoting EDIFACT to the electronics sector.

12.8 BENEFITS OF EDI IN MANUFACTURING

12.8.1 Paperwork

As with most other industries, the initial benefit of EDI in a manufacturing industry is a reduction in the cost and time taken to generate and process orders, invoices, delivery notes, and schedules. In a large business, there can be significant savings in manpower; small businesses are less likely to reduce the number of employees, but will probably find that the same number can handle a larger workload or give better service to customers and other departments.

The cost of processing a purchase order or similar document varies from industry to industry and from company to company. These documents are generally more complex for industrial goods manufacturers than for others. Once you add up the time taken to open an order, check it and correct any errors, resolve any ambiguities, enter it into the computer system and confirm it, a figure of $15 can generally be taken as a minimum, and $25 to $50 is more common. EDI users report that these costs can be reduced by more than 50%. Table 12.2 shows the experience of Philips UK in reducing the cost of handling incoming and outgoing documents. Digital and Hewlett-Packard have experienced similar savings.

Since EDI is often implemented alongside a production control system and provides additional and more timely information to that system, there is often a large reduction in the time taken to answer customers' progress calls. In fact, such calls may become unnecessary, as they can be made by EDI. This should be the major benefit for manufacturing industry in the new interactive EDI standards.

12.8.2 Communications Costs

With EDI, big savings can be made in communications costs. While the EDI messages are transmitted over a telephone line and therefore incur a cost, this form of data transmission is

Table 12.2
Cost Savings at Philips UK

Cost (£/1,000 documents)	Manual	EDI	Savings
Incoming	1,510	325	78%
Outgoing	110	55	50%

Note: Reproduced with permission of Philips.

much more efficient than the telephone or even the fax. Because the data is generally only transmitted once, there is little waste or duplication. A key factor is that the data exchange with EDI is managed. Noncritical data is handled during off-peak hours in batch mode, while more urgent messages can be transmitted immediately.

12.8.3 Errors

Errors in the paperwork associated with a manufacturing operation can be very costly. They may result in the wrong part being manufactured, the wrong materials being ordered, or simply payment or manufacturing delays. Many companies find that at least half their sales administration time is taken up correcting errors on orders or chasing details on incomplete orders. With EDI, where the original data will have come not from a keying operation but from the trading partner's computer, these errors should be much less frequent, and can with time be eliminated.

12.8.4 Design and Product Data

EDI is not the main way of exchanging design and product data between organizations today. With higher-speed digital links, however, this should change. As described in Chapter 7, standards for this type of exchange are being developed, and they should be available for use within the same time scale as the telecommunications links.

When this facility is established, however, the benefits will be considerable: design data should be more up-to-date and more accurate, while the problems of exchanging data between different CAD systems should be largely eliminated. The product data standards such as STEP and EDIF aim to extend far beyond the mechanical drawing or circuit diagram, and a fully coordinated production process should be possible using these standards before the end of the century.

12.8.5 Lead Times

The lead times required for deliveries in typical manufacturing situations can be greatly reduced through the use of EDI. Some improvement can be obtained immediately through the elimination of postal delays, errors and order processing time. The larger part of the improvement depends, however, on the purchaser supplying reliable production schedule data and on the supplier using the information supplied. This implies a high level of trust between the two companies and often a different form of commercial relationship. Single or dual sourcing agreements are much more likely to be explicit than in a paper trading environment.

12.8.6 Inventory Savings

For manufacturing industry, the greatest potential savings from using EDI come from reducing inventory and the cost of holding this inventory. We have seen that reductions of 80% and more are possible. The savings may be even greater than this: it should be possible to reduce stock write-offs (usually the result of over-production). The fact that goods are always produced very close to the time they are required should also cause fewer quality problems: each person in the chain feels closer to the end product.

12.8.7 Process Redesign

The cumulative effect of all the savings described above would be more than enough to justify the introduction of EDI in the majority of manufacturing operations. But introducing EDI does not in itself guarantee any savings or benefits whatsoever. The manufacturing process must be redesigned to exploit the information available and to provide the data required by suppliers and customers. It is really the process redesign made possible by EDI that delivers the savings Any reorientation of the manufacturing process to concentrate on more productive goals will of course be beneficial. But a redesign that takes advantage of all the information available and that allows production to control the flow of materials and resources is the most advantageous.

One often overlooked benefit of such process redesign is the elimination of the "tyranny of administration." This is the feeling that production is working for the administration: filling in forms and recording details more than actually making things.

12.8.8 Customer Service

Any organization that wishes to introduce EDI will almost certainly have to justify it in terms of savings. There may be room somewhere, however, for a short word on behalf of the customer. The combination of EDI with a just-in-time or other demand-driven production system should result in vast improvements in the service offered to customers. It should be possible to give the customer more accurate delivery estimates and progress information. Because products are made to order rather than drawn from stock, different variants or options can easily be incorporated. And as we said earlier in this chapter, this should not lead to any increase in delivery times. Customers who are themselves EDI users will of course benefit most: they too will save on paperwork and communications costs, and on time correcting errors. But the biggest benefit lies in the closer relationship with the customer and the mutual commitment that an EDI partnership involves.

Chapter 13
Electronic Funds Transfer

13.1 THE BANKING INDUSTRY IN EUROPE

The structure of banking varies considerably from country to country. While the most obvious differences lie in the structure of bank branches and the ownership of banks, there are also important differences in the payment methods most commonly used. Countries can be roughly divided into the "debit-driven," or check-based countries, and the "giro" countries, where credit instruments are more common. Beyond these differences lie a host of social and political attitudes to banking that will continue to slow progress towards any form of international homogeneity.

In most of the major European countries there are two or three major banks that tower above the others in their sector: Crédit Lyonnais and Société Générale in France; Deutsche and Dresdner (with Commerzbank just behind) in Germany; National Westminster, Barclays, Midland, and Lloyds in the United Kingdom. Behind these dominant players stand a dozen or more smaller commercial banks followed by the more-specialized banks: savings banks, merchant banks and investment banks. We are here concerned mostly with those banks that are involved in the regular transmission of funds on behalf of their customers through what is known as the clearing system. The clearing system is the mechanism by which the banks exchange funds that they then credit to or draw from their clients' accounts.

The banking industry in Europe was slow to start the move from traditional banking, where mahogany panels and marble floors were considered essential to a bank's image of stability and dependability. A dramatic change occurred, however, when cash-dispensing machines (the predecessors of today's Automated Teller Machines—ATMs) were introduced in 1967. Soon there were queues in the rain outside for the ATM, while the inside of the bank was empty. The size of a bank's ATM network became an important competitive factor. There are now more than 100 major ATM networks in Europe, with a total of over 50,000 machines. Most of these European ATM networks are linked to international networks.

Table 13.1
ATM Networks in Major European Countries (1990)

Country	ATMs	ATM networks
Belgium	913	2
France	13,031	1
Germany	9,300	4
Italy	7,791	1
Netherlands	1,839	2
Sweden	1,794	2
Switzerland	1,962	2
United Kingdom	15,820	21
Totals	51,450	35

Once the move had started, the banks' appetite for telecommunications grew at an amazing rate. This was particularly true for London, where "Big Bang" in 1983 caused strong competition from American banks. This is now also true of Frankfurt, Paris, and Zürich, all of which are competing for London's place as the preeminent financial center in Europe.

Telecommunications' influence was most widely felt in three areas. Telecommunications were essential for (1) linking the banks' branches with their head offices, (2) keeping the bank up to date with market and international developments, and (3) making the payment system more efficient.

13.2 MONEY TRANSMISSION AND CLEARING SYSTEMS

Figure 13.1 shows the traditional "clearing" mechanism used for payments between two trading partners. Able Baker Limited issues an instruction to its bank to transfer money to Charlie Dough and Company's account, which is with a different bank. The banks exchange this information on Monday and on, say, Wednesday of that same week Abe Baker's account is debited with the money while Charlie Dough's account is credited.

On Wednesday evening, the two banks add up all the transactions between them, and whichever one owes the other pays the difference. This process is called netting. The banks usually aim to do the debit a few hours before the credit; the interest on this one transaction may not amount to much, but, when all the day's transactions are added together, it may make the difference between lending and borrowing in the money market. This temporary balance is called the float.

When the two banks are located in different countries, the picture becomes only slightly more complicated, as shown in Figure 13.2. International payments frequently involve the intervention of a "correspondent bank" with whom the paying bank has a

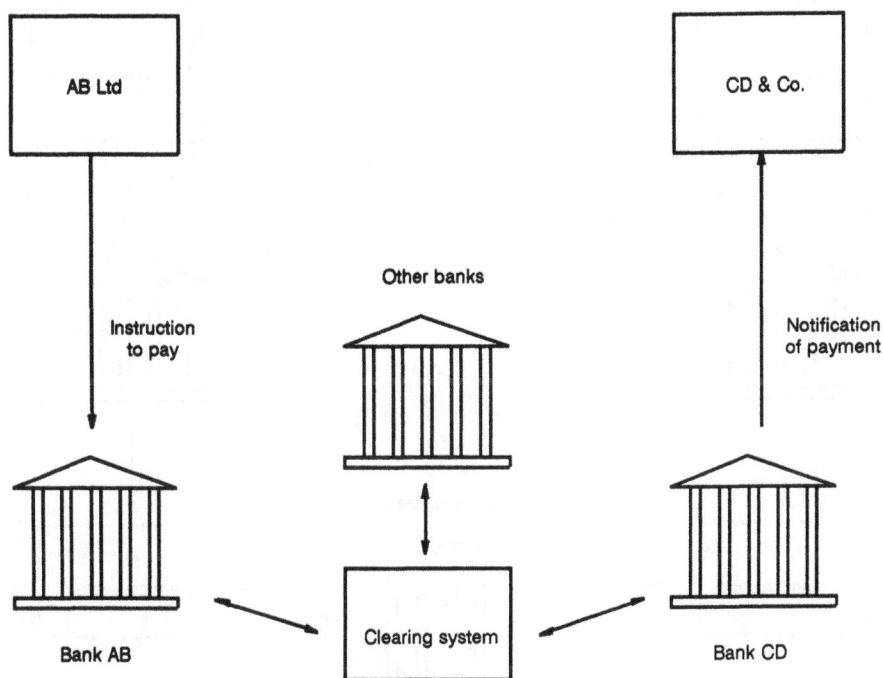

Figure 13.1 Domestic clearing.

contractual arrangement. Exchange rates and different financial systems tend to obscure the transaction further, and introduce an element of uncertainty that discourages companies from trading internationally.

The first EDI systems in banks were part of the clearing system itself. At first, tapes were sent, then fixed links were set up between the banks. This meant that the transaction details only had to be keyed in once, although in many countries it is still a legal requirement that a paper "payment instrument" (such as a check or giro form) is also sent to the bank on which it is drawn.

Today, checks, giro slips, and other paper instruments are converted into electronic form as quickly as possible. Optical character readers (OCRs) and magnetic ink readers are used to capture the account details from the form, and the additional details are entered as soon as the form enters the payment system.

With a computerized clearing system, it is relatively easy to "net" transactions: that is, we subtract all the money I owe you from the money you owe me, and you only pay me the difference. This reduces the total transaction costs and makes the system as a whole more efficient. Almost all transfers between clearing banks in Europe, both

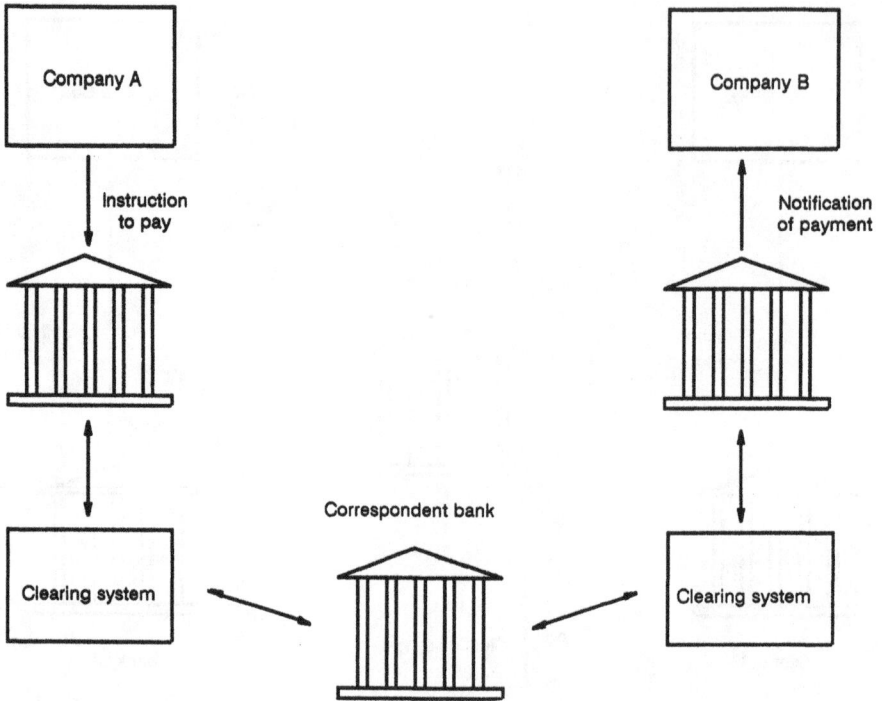

Figure 13.2 International clearing.

nationally and internationally, are carried out electronically—this is referred to as interbank EDI. In the United States, the Banking Acts have up until now limited this process, but ways are now being found to extend interbank EDI to a larger proportion of the smaller banks.

In the 1970s, several countries introduced automated clearing houses (such as BACS in the UK) that allow traders to send bulk payments, such as a payroll or regular payments to suppliers, directly by magnetic tape. As these traders became EDI users, they sought to make their payments by EDI, and this is now becoming possible through a wide variety of new schemes. The use of EDI for trade payments is, as we shall see, a major growth area.

In retail banking, card payment systems are the biggest growth area. As seen in Figure 13.3, payments by credit card, debit card, and charge card have already reached over 20% of all payments in retail outlets in both France and the United Kingdom. Although this is still some way behind North America, these countries have a substantial lead over the other European countries, where more expensive (but also more secure) systems have been put in place for card payments.

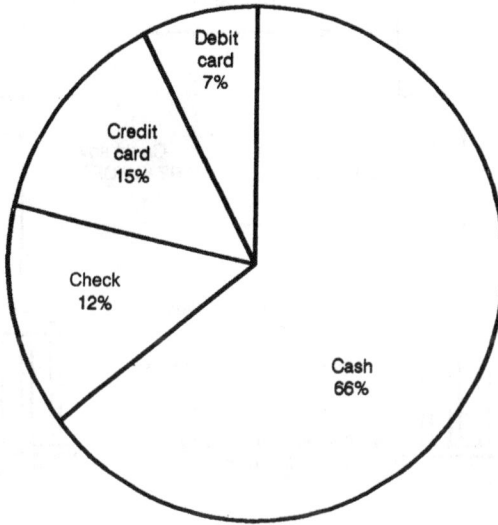

Figure 13.3 Distribution of retail payments in the UK (1991). Source: The author's estimates from APACS and CSO data.

13.3 INTERBANK EDI PROJECTS

The first systems used proprietary, fixed-format message standards and private networks or fixed links. As the number of participants grew in the 1970s, however, it became clear that this would constitute an excessive overhead, and the move towards relatively open systems became irreversible.

In most EDI systems, customers are concerned about the security of the whole system and need reassurances that maximum precautions have been taken against wholesale breaches of security. But individual transactions are rarely regarded as highly critical from a security point of view. Maximum transaction values are often set, and companies knowingly take the risk within these limits. In financial EDI, however, this is no longer true. The level of security demanded at the individual transaction level is much greater than elsewhere in EDI, since millions could be at stake. Great reliance is placed on physical security at terminals, and access control and authorization levels are carefully controlled within the protocols.

Using the EDIFACT messages, Figure 13.4 shows how the clearing system described earlier in this chapter is translated into an EDI system. One important difference between the conventional system and the EDI systems implemented to date is that settlement in

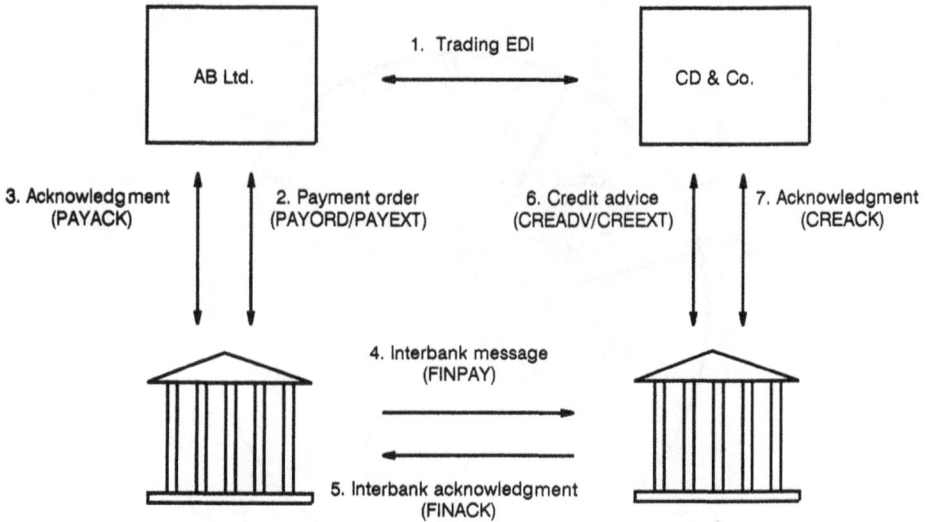

Figure 13.4 Domestic transaction using EDI.

the EDI systems is bilateral; that is, each bank settles with every other bank (after netting) rather than through a central clearing service. This is partly because clearing is much easier with EDI but also because interbank EDI is not yet sufficiently developed or standardized for any of the banks to trust an independent body to carry out clearing.

Many national standards are used for EDI between banks. When the systems include direct message exchange with customers, however, the EDIFACT standards and UNSMs are almost universally used. Financial EDI will only work and become accepted if the service is truly international.

The most important messages in Figure 13.4 have now reached Status 2 (Approved Messages). A further set of messages relating to direct debits (the most common form of trade payment in France and Italy in particular) should be approved by 1996.

Trials of an EDI system for documentary credits (frequently used in international trade) have recently started in the United States, but this is unlikely to be a major application for some years. Documentary credits involve a very large number of players, often including as many as six or eight banks, and it will take some time to reach a level of penetration where such a system can be widely used.

Some of the most important projects in interbank EDI are described below.

13.3.1 SWIFT

SWIFT is one of the earliest, and still one of the most important, applications of EDI in the financial sector. It is not an exaggeration to say that nearly all routine operations in international banking make use of it. The Society for Worldwide Interbank Financial Telecommunications (SWIFT) was formed in 1973. Based in Belgium, it has around 1,500 member banks in 75 countries, and is the main channel for international interbank clearing; that is, matching the two halves of a transaction where the accounts or currencies in question are located in different countries.

SWIFT has its own proprietary network (which is currently being upgraded) and uses its own fixed format message standards. The following messages can be transmitted by SWIFT:

- Common group;
- Customer transfer;
- Bank transfer;
- Foreign exchange, loans and deposits;
- Collections;
- Securities;
- Syndications;
- Documentary credits and guarantees;
- Travellers checks;
- Special messages;
- Service messages.

Since 1991, SWIFT has offered a new "true" EDI service based on the EDIFACT PAYORD and PAYEXT messages but slightly modified because of limitations on the international network. Initial takeup of this system has been slow, with only around 250 messages a month at the end of its first year, but most of the participating banks believe in its future. Other new systems, such as a new bulk payment standard, continue to be developed using the traditional SWIFT syntax and protocols. SWIFT found that the long approval procedures for EDIFACT would have made it difficult for it to offer all the services that its customers needed.

13.3.2 United Kingdom

Interbank EDI has a long history in the UK. As elsewhere, the first systems were services operated primarily for interbank clearing and using magnetic tape input. Although these were gradually extended to allow direct customer access and online links, they are regarded as the precursors of today's EDI systems.

The Bankers' Automated Clearing Service (BACS) was set up in its present form in 1985. It is owned by the major UK clearing banks. BACS operates on a central mainframe system, to which all the clearing banks have direct links. Clearing (sorting and matching transactions to each of the banks) takes place at the end of each working day. Large customers can also send tapes or dial in to BACS using EDI-like formats and protocols. PC packages now make this a relatively accessible payment method. There is, however, little scope for sending references or other information to accompany the data.

The Clearing House Automated Payments System, described in Chapter 1, is closer to a conventional EDI system. CHAPS has its own closed user group on a digital (packet switched) network. Banks use CHAPS to send guaranteed sterling payments to one another; only they can access CHAPS directly, although customers can request a CHAPS payment.

Since 1988 there has been rapid progress towards offering EDI directly to customers, as described below. Most of the banks now offer such a service. The next stage is already being developed through the Association for Payment and Clearing Services (APACS), a body owned by the banks. The Interbank Data Exchange (IDX) network aims to provide seamless transfer of EDI payments, credits, and remittance advices between UK banks on behalf of EDI traders.

13.3.3 France

Parallel with the considerable penetration of digital networks in France during the 1980s a network for interbank payments, ETEBAC, has grown and virtually replaced the previous paper and tape-based systems.

The *Systéme Interbancaire de Télécompensation* (SIT) was set up on the ETEBAC network in 1989. This is an EDI system using proprietary message formats and syntax. SIT is capable of being extended to provide a full EDI service to traders.

In France, letters of credit are frequently used for domestic trade payments, whereas elsewhere in Europe these are mostly used in international trade. Letters of credit require documents to prove, for example, delivery before a payment is made. This need for matching documents in order to complete a transaction changes the emphasis within the EDI system.

13.3.4 Other European Countries

The SIA (*Societa Interbancaria per l'Automazione*) system in Italy and SITO (*Sistema Interbancario de Transmision de Operaciones*) in Spain are similar in principle to France's SIT.

Several European banks have joined in a service called Ebic. This uses the SWIFT EDIFACT messages for the payment aspect and the IBM EDI network Information Exchange for the associated trade messaging, including remittance advices.

13.4 COMMERCIAL BANKING EDI

Whereas EDI, in various forms, has been a central part of interbank operations for many years, it is only recently and in one or two countries that banks have started to allow their customers to make financial transfers by this method. Several banks now offer packages to customers to allow them to format data and instructions in a form compatible with the banks' systems. Electronic banking is now offered not only to personal customers (via interactive packages such as the French Minitel service) but also to business customers.

In the United Kingdom in 1988, a set of EDI trials was conducted in order to find out what kind of service the bank's business customers wanted; the most important of these trials formed the payment link in an EDI system operated by the car manufacturer Peugeot Talbot and one of its large suppliers, Lucas Industries. By 1990, most of the major UK banks had available a commercial EDI service.

Customers are now offered a range of software packages complete with security tools such as smart cards. These are usually for PCs, although the standards can also be readily handled by any conventional EDI software running on other hardware platforms. The range of software and services available gives customers confidence that they are not becoming "locked in" to a bank through technology. These are described in more detail in Chapter 15.

So far in Europe, no bank has yet adopted the approach of the Royal Bank of Canada (RBC), which has developed its own EDI network for its corporate customers. The R*EDI*CON service offers exchange of trading messages as well as electronic funds transfer.

In the UK, many different networks are used for the collection of transaction data from retail EFT-POS (electronic funds transfer at the point of sale) terminals. These data can also be integrated with retailers' other EDI messages and carried over the same network. This is, however, unusual, as in most countries the card payment network is kept separate from other VANs.

An important point to bear in mind is that a company can use EDI for electronic payment even if it is not using any other form of EDI. Most of the commercial banking EDI packages will interface with common PC accounting packages, so the only requirement is that the company uses a computerized accounts system. Some of the banking services include generating and sending out printed notifications, while other services (such as the Royal Mail's EDIPOST service in the UK) will deliver a printed form directly from a standard EDIFACT message when an EDI user's trading partner is not able to accept the EDI message.

13.5 RETAIL BANKING EDI

13.5.1 Credit Card Clearing

The worldwide credit card system is a vast EDI network. Within each country, transactions are collected from retailers, using either the public switched network or a special card-payment network. The banks that are authorized to collect these transactions (known as merchant acquirers) then sort and distribute them to the other banks according to the rules of the various card schemes (e.g., Visa, Eurocard/Mastercard).

Debit cards such as Dankort in Denmark and Switch in the UK sometimes offer a slightly higher level of security, but, for credit cards, it is still common to use the public switched network and a published standard, with little or no encryption, for data transfer. This is a remarkably open standard for financial transactions and constitutes a significant security risk. Steps are only now being taken in some countries to minimize this risk.

Card payments are often associated with credit cards, particularly the Visa and Mastercard/Eurocard schemes. Although in several countries, credit cards are the norm, there is evidence that it is the convenience rather than the credit that attracts users. In other countries, debit cards (where the payment is taken directly from the customer's account) are more common, or credit is only given for a fixed period.

Whereas the first credit card systems relied on paper slips that were data-captured at a processing center, there is now a strong trend towards using electronic terminals at the point of sale. Many retailers use an electronic register with card-reading facilities linked to a back office system that may also do stock control and end-of-day procedures; this is known as an electronic-point-of-sale (EPOS) terminal. Others have special payment terminals that can read the magnetic stripe or smart cards. The shop assistant then enters the amount, and the customer either signs or enters a personal identification number (PIN).

In an offline system, as operated in France and the UK, most transactions are simply stored in the terminal, which is then polled by a central computer at night. In the German and Belgian systems, every transaction requires the terminal to connect online to the central computer for authorization and immediate clearance. Other countries have generally adopted a combination of the two.

In either case, the traditional clearing system is made more complicated because the retailer accepts many different cards issued by many different banks. These transactions are "acquired" into the banking system by a merchant acquirer who, in most cases, must also be a bank, but need not be the retailer's bank. Other parties such as a polling bureau or independent card processing company may also be involved.

When electronic card payment systems were first introduced, proprietary standards were used for data transmission. In the UK, the Association for Payment and Clearing Services (APACS) first defined standards for authorization and transaction handling messages; these are known as APACS 30, 40, and 50 and are also used in many other countries, including Scandinavia. In France, the GIE Carte Bancaire (the grouping of Visa banks) defined the standards that are used in that country, while in Germany and Belgium the ISO standard for interbank transaction transmission is used.

A set of EDIFACT messages for credit card authorization and transaction handling has also been proposed, but this has not yet been tested. None of the standards currently in use has achieved a satisfactory level of security coupled with the efficiency needed for the tens of millions of transactions that are processed every day; this area is still being developed. Many of the systems and underlying banking laws differ significantly from country to country, and it will be many years before a fully international electronic scheme will be operational.

13.5.2 Home Banking

Personal customers in some countries can now check their balances, make payments, and conduct other banking transactions through online terminals in their homes. These facilities are available, for example, in France through the Minitel network and in the UK through the Royal Bank of Scotland's HOBS service. German customers have more limited services available through the Btx (*Bildschirmtext*) service.

The main limitation on the expansion of these services is the small number of households who have true home computers (as against portable office machines or games computers). It is significant that the services in France, the UK, and Germany all use videotex terminals. Videotex (or viewdata) was thought in the 1970s to be the obvious route to home computing, but in the end it was found that terminals and modems were either too slow or too expensive for normal home use. It will be interesting to see whether the growth of ISDN revives the videotex market and with it home banking.

13.6 BENEFITS OF ELECTRONIC FUNDS TRANSFER

It is often argued that the benefits of electronic payment are very one-sided: one party gets paid faster, and the other has to pay faster. The supplier is therefore keen on using EDI for payments, but his customer is less keen. This argument arises when the two companies have not considered the true benefit to both of them. They can, in fact, choose to carry out the payment on any day: the day after the invoice is raised, thirty, sixty, or ninety days. The EDI system does not of itself dictate the terms of trade. What it does do is to make it possible to pay the day after the invoice (even the same day as the invoice) and to reduce greatly the cost of making the payment. For transactions of less than a few thousand dollars, the cost of the paperwork is greater than thirty days' interest on the money.

Corporate treasurers often have the major say in the implementation of a new payment system. They are often more conscious of the cost of money and bank charges than of the cost of personnel, while it is often an operational department that will reap the benefits. Agreeing on the mutual benefits with a supplier may fall to a marketing or even a computer services department. People are more conscious of the cost of the paperwork and extra bank charges in international payments, so it may often be easier to persuade a treasurer of the benefits of EDI in this case.

For the banks, the advantages of EDI are compelling.

- Only with EDI can the process of payment transmission automation (and a further lowering of the service costs) be completed. All the standard benefits of EDI, including speed and reduction of errors and paperwork, apply doubly to the banks because of the very large volumes of transactions they handle.
- Large customers are moving to EDI and are demanding electronic payment as a part of their banking service. Every sector that starts to use EDI wants to complete the loop by adding in the payment mechanism.
- It is unusual for the banks to be able to offer a new banking product that has such significant added benefits. The banks are therefore eagerly using EDI as a marketing tool for acquiring new business.

Many banks are concerned at the greater degree of transparency that EDI gives to their operations. Customers are more aware of the time delays and float attributable to the banks. The instant and timed acknowledgement offered in EDI payment imposes a higher quality standard. In fact, this can be turned into an advantage for the banks as well as their customers, as raising awareness of problems is often the best way to start resolving them and offering a better service. As we will discuss in Chapter 18, EDI is fast becoming a necessity rather than a marketing tool.

A more real danger for the banks is that EDI could cut them out of the loop in some banking transactions. In the same way that banks use a central clearing system and net transactions before making a payment, so a group of traders could also. Banks have already lost their monopoly on several other services that do not actually require a banking license (such as credit references). EDI could be the first real threat to one of their core businesses: payments transmission.

At the very least, businesses will seek the best possible service and value for money from their EDI payments system, and this will become a very competitive area within the banking field. The other side of this is that businesses will be able, as a result of competition, to demand and obtain the best service at the best price. Customer loyalty in an internationally standardized service will be very much less than in traditional banking, and banks with a low cost base (i.e., low domestic and international communications costs) will have a major advantage.

Chapter 14
Other Applications

14.1 REQUIREMENTS FOR EDI

An industry sector must meet certain qualifications before it is able to make use of EDI. An individual company within an industry can of course make use of any of the applications we have already described. As financial EDI develops, larger or more automated companies are likely to move to using electronic payment even if only a few of their customers are doing so, and certainly regardless of other companies in their own sector. For an individual company to use EDI, the only requirements are that:

- It uses a computer system for the whole of an application (such as ordering, invoicing, accounts, or computer-aided design).
- This application has an input that comes from another company (such as an order) or an output (such as an invoice) that is sent to another company, and this other company in turn generates the order or records the invoice also on a computer.

For us to benefit from EDI, it is not necessary for all our trading partners to be computerized, but, at a minimum, one or two key customers or suppliers should be able to use or generate the information. The more the better. A few of the advantages of EDI can be obtained from developing a system with just one or two key trading partners: for example, a manufacturer whose main customer is a large supermarket chain may develop a production control application centered around the EDI orders coming from that customer. But this will almost certainly be to the disadvantage of all his other customers, and one of the main advantages of EDI will be lost: the ability to streamline or redesign the whole production process based on order inputs.

For an application to grow within a sector, several further conditions must be met.

- There must be a generally accepted sequence of doing business in the sector: one sector may expect suppliers to prequalify (company and product appraisal), then submit sealed bids that are evaluated by a purchasing department, following which a single order is placed. In another sector, local managers may place orders based on their knowledge of competing products and delivery times. Or a "blanket order" may be placed against which local managers may call off specified quantities.

- There must be a common terminology. In many sectors, suppliers frequently say that "our CM234A is the equivalent of Smithsons' XY777." Electrical cables and integrated circuits are cases in point. Without this equivalence or common terminology, an industry-wide application cannot be implemented. Airlines must all use the same codes for a given airport, and banks the same codes for currencies.
- EDI applications are most likely to grow where there is a moderate number of larger firms within a sector. Where one or two companies completely dominate, competitive pressures will prevent them from making an application available to other firms. An industry that is made up entirely of small organizations will find it difficult to coordinate and to devote the effort to the creation of such systems.

Some examples of sectors that have been able to develop industry-wide applications are given below.

14.2 INSURANCE

The insurance industry is not traditionally a major user of telecommunications. In recent years, however, insurance companies have become much keener on technology, and it is now common for brokers or insurance company representatives to work out premiums or to draw up comparative policies using computers. The premium data on these computers must be updated from time to time, and this often demands an increasing use of telecommunications. A large amount of routine paperwork is also exchanged between insurance companies and their brokers and branch offices. This is eminently suited to an EDI approach, and a number of companies in Europe are now starting to adopt such a principle.

In France, a network called CELIAS is used for sending monthly premiums and remittance advices between brokers and insurance companies. In the UK, an EDI system called LIMNET (London Insurance Market Network) is used to connect brokers, underwriters, Lloyd's, and the major insurance companies. LIMNET uses the EDIFACT syntax together with some of its own message definitions. Most of the 1,100 Lloyds underwriters, their agents, brokers, and commercial insurance companies will eventually be connected to this service. An Edinburgh-based insurance industry organization called Origo is introducing standard EDI messages for proposals and commission charges. Although there is a relatively fixed way of selling life insurance policies through intermediaries, Origo has found that there are enough difficulties between the ways its members do business to ensure a rather slow start to the service. Reinsurance (where insurers limit their risk by placing further policies with other companies) is the most international part of the insurance market. A European network called RINET (Reinsurance and Insurance Network), serving the reinsurance industry, is currently under development.

14.3 SECURITIES

Share certificates should be one of the most standardized pieces of commercial paper and hence the easiest to exchange by computer. In practice, although many companies operate computerized registers, legal and historical difficulties abound in the creation of computerized stock and share exchanges. For companies traded on different exchanges, this is one of the most difficult areas to coordinate because of the wide variety of different systems used by the world's stock exchanges. Many stock exchanges, including London, have had extreme difficulty just in computerizing share registration; if this experience is anything to go by, exchange of data between exchanges is several decades away.

14.4 TRAVEL AND TICKETING

Companies in the travel industry for many years have been used to accessing each others' computers to check schedules, availability and prices, and to make bookings. The airlines' online reservation systems, such as Sabre, Galileo, and Amadeus, are among the most extensive and sophisticated systems of this type in the world.

For more general use within the travel industry, however, a less elaborate system is required. In North America, direct access to a wide variety of different operators' booking systems has been the norm, but, in Europe, videotex has been the traditional vehicle. Either approach makes it very difficult for travel agents and others making bookings to interface smoothly with their internal systems, and errors are common.

A form of EDI messaging is clearly the answer. Normal EDI, however, deliberately "uncouples" the input and output processes and performs a buffering role. This is not acceptable in the travel reservation environment, where very often the customer is waiting in the shop for a response. What is needed is a form of interactive EDI, and this has led to the I-EDI initiatives described in Chapter 7 and to the development of event-driven EDI software.

The same principle applies to other forms of ticketing and reservation systems. One of the first projects to use these principles was UNICORN (United Nations EDI for Co-operation in Reservation Networks), which links together the operators of North Sea ferries. UNICORN started with its own modifications to standard GTDI messages, but is moving to EDIFACT standards.

14.5 OTHERS

Within the construction and building-supplies industry, EDI may be used not only for ordering and invoicing, but also for bills of quantity, product information, estimating, and project control. In the UK a construction EDI group called EDICON (EDI Construction, Ltd.) was formed in 1987.

There are several EDI projects in the electronics sector. Most of these involve only a small number of companies, partly due to the difficulty of agreeing on standards in this fast-moving area. A European special interest group called EDIFICE (Electronic Data Interchange for Companies with Interests in Computing and Electronics) aims to promote the use of EDI in electronics, mainly in business and financial applications. Electronics companies in North America are more active in the area of technical EDI for the exchange of CAD and product data.

In the oil industry, there is an EDI system for accounting for fuel sales and fuel exchanges, including the large quantities of data relating to volumes and discrepancies, duty and VAT, and other handling charges. European oil companies are now starting to participate in a North American system, Avnet, for aviation-fuel orders and invoices, which must also pass internationally.

Other special industry groups within the EDIFACT framework exist for accountancy and audit, clothing and textiles, and the paper industry.

Many other EDI projects are variations of buying and selling. Earlier in the book, we examined the case of schools using EDI to consolidate their orders for routine supplies in order to place them through a central purchasing system. Grocery companies have been doing this for years; well-known organizations such as Spar have grown up entirely on this principle.

In Chapter 7, we mentioned a number of technical EDI applications, in particular the transmission of product data (drawings, specifications, maintenance instructions). These are still in their infancy, but, when they become integrated with commercial EDI systems, will add greatly to the power and benefits of EDI.

14.6 FUTURE APPLICATIONS

Future applications of EDI are limited only by the imagination. Increasing numbers of networks are transaction-driven, and, with the advent of interactive EDI, the concept can be extended even further.

Starting with commercial and financial EDI, there is already provision for a range of transactions far wider than most companies are able to use. While a high proportion of companies already use computerized invoicing or production-control systems, very few trading companies use computer systems for documentary credits, credit referencing, or forward currency purchases. In these cases, the standard has evolved ahead of the need.

Lagging behind are areas in which tapes or floppy disks are regularly used for data exchange. These include bulk payments such as payroll systems, taxes, and direct debits for utilities (water, gas, electricity). Here, the standards exist, but, although the change to an EDI system would bring cost savings, there is little or no pressure from the trading partner. Floppy disks are often used for updating files held remotely: price lists, insurance rates, stock lists, and the like. Very often, these remote systems do connect with the central system from time to time, but floppy disks are a more economical means of transmitting the data.

Two groups of organizations aim to change this: (1) telecommunications and value-added network operators, who aim to increase the speed and cost-effectiveness of their networks to the point that all data can economically be carried over them; and (2) satellite and other data broadcasters, who can transmit the same data to many locations very quickly and at no greater cost than transmitting it to a single location. This technique is underutilized partly because EDI formats have not yet been widely used and each application therefore developed separately. EDI would enable data broadcast to interface with standard packages using an open systems concept. Figure 14.1 shows how a data broadcast network operates. Data broadcast is currently used for stock market prices and other financial data, news flashes and news gathering systems, betting, credit card "hot lists," and broadcast messaging. All of these are candidates for EDI in a broadcast mode.

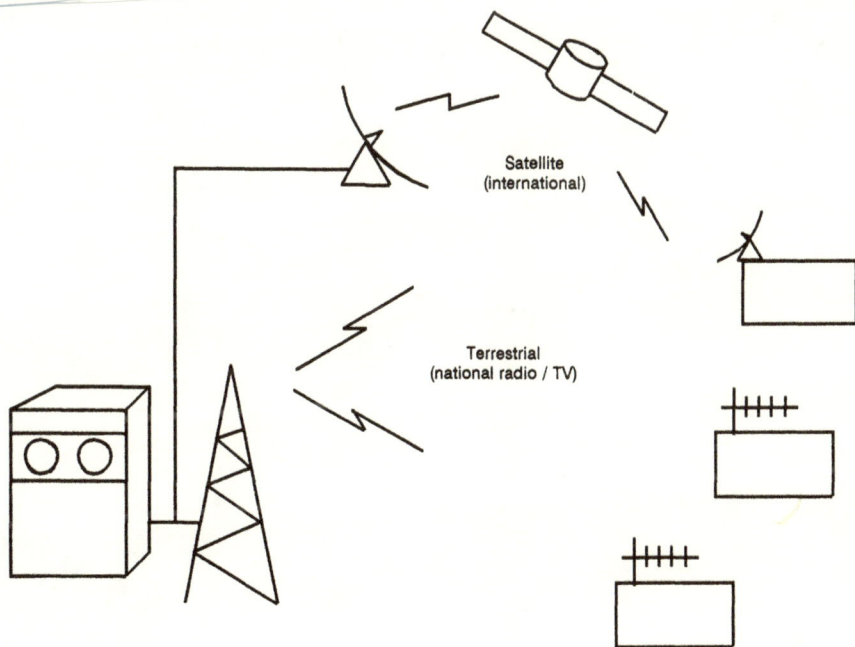

Figure 14.1 Data broadcast network.

Chapter 15
EDI Services and Networks

15.1 TYPES OF SERVICE

As we discussed in Chapters 5 and 6, one can make use of EDI without any special hardware, software, or network—what might be called the "do-it-yourself" approach. In practice, the hardware is always fairly standard, and software for translation of your data formats into a recognized EDI protocol is readily available from several competing vendors. Although these software packages can and do vary in their power and user-friendliness, most of them can cope with a fairly wide variety of different EDI protocols and communications standards, including EDIFACT, TDI or its local variants, and ANSI X.12.

Many users, however, are faced with difficult decisions when it comes to choosing an EDI network or service. For a start, one needs to decide what people mean by an EDI network: Is this simply a network that will transfer data from one node to another, or does it have features specific to an EDI application? Most users feel that an EDI network should, at a minimum, have store and forward facilities and probably some auditing and record-keeping capabilities as well. Many will also perform some checking on messages received to ensure that they can be read by the receiving party. An "EDI service" will go further than this. It should have suitable access control checks, keep a profile of trading partners, know who can send or read what message types, and possibly also provide other value-added functions such as broadcast messaging, statistical reporting, or transaction clearing. Many EDI services are specific to one or more industry sectors.

For many EDI operations, however, none of these facilities are necessary. Two companies that wish to exchange data can agree on simple manual or automatic procedures and call each other once a day. The increasing complexity and power of public data services can be used directly for exchanging EDI messages. With worldwide X.400 messaging coming along, many EDI networks could indeed become redundant. For the moment, however, value-added networks are still the main providers of EDI services. Most of the major VANs have at least one service operating on their network.

15.2 MAJOR EDI NETWORKS

There is a fairly common pattern among the major EDI networks: they normally provide electronic mail (e-mail) facilities as well as EDI and often other forms of data transfer or interactive access to databases. Most can support asynchronous or synchronous (2780/3780) access, X.25 and X.400 (with either P1 or X.435 standards for EDI messages), and either ODETTE (in Europe) or AIAG (in North America) file transfer standards. As far as EDI protocols are concerned, all will support EDIFACT and ANSI X.12, usually with some older or sector-specific standards relevant to their historical business. What follows is a brief description of the four largest EDI networks in Europe and North America.

15.2.1 AT&T EasyLink

AT&T EasyLink was formed in 1991 and brings together AT&T's Global Messaging Services and Western Union's EasyLink messaging service. It provides electronic mail and EDI, as well as store and forward fax and telex services that can be used by EDI users to send documents to trading partners who are not equipped for EDI. The EDI element includes both AT&T EDI and EDICT, an EDI system developed by Istel in the UK, which is now an AT&T subsidiary. AT&T in the UK operates as AT&T Istel, since Istel had an established customer base, particularly in the travel and electronic payment sectors. The worldwide network is linked through nodes in the US, UK, and Japan. The network uses a message transfer architecture that meets X.400 standards and allows a single connection to transfer EDI messages, electronic mail, and files of any size. It is thus particularly suitable for users whose requirements extend to technical as well as commercial EDI. The software offered is Istel's EDICT-PC or other third-party packages.

15.2.2 BT Tymnet

British Telecom operates an EDI network on its Global Network Service (GNS). The service uses the EDI*Net system developed in the United States as a joint venture with McDonnell Douglas, and fully acquired by British Telecom in 1989. All of the major EDI standards are supported, and access can be by dial-up, X.25, or 2780/3780 synchronous protocols. GNS also offers a direct point-to-point packet-switched service and an X.400 messaging service.

15.2.3 GEIS and INS

General Electric Information Services (GEIS) is one of the largest networks worldwide. A significant EDI player in its own right, GEIS also owns (with computer manufacturer ICL) the largest EDI network in Europe—International Network Services. GEIS has offered its EDI*EXPRESS service since 1980. It covers a very wide range of sectors, and has connections into many other networks. All of the usual protocols and access methods are supported; there is also an X.400 messaging service on the same network.

INS operates entirely within the United Kingdom and runs the main retail EDI service, Tradanet, as well as several other sector-specific services. INS can handle all the main EDI standards, although Tradanet uses primarily the Tradacoms standard, a version of the older UN/TDI standards. INS users are offered the INTERCEPT software package appropriate to their hardware: MVS, AS/400, or PC; in the case of PC users, this is included in their network joining fee. INS has a seamless link to GEIS for international operations as well as connections to AT&T EasyLink. Transnet in the Netherlands is also a part of the group. INS has over 5,000 EDI users; it is more difficult to identify GEIS' EDI users from the 13,500 on its network. Between them, the two services handled over 120 million messages in 1991.

15.2.4 IBM Information Network

IBM's Information Network is slightly different from the other EDI networks described here, in that it is more or less independent of protocols or EDI standards. Its function is to transmit data between any two companies on the network—although if any of the standard protocols are used, then they can be checked. IBM Information Network has over 50,000 users, but relatively few of these make use of the EDI services. Those who do, though, are often heavy users, including IBM itself.

Information Exchange is the EDI service offered on the Information Network (see Figure 15.1). The European node for this network is in Warwick, England, while the other nodes are in Tampa, Florida and Tokyo. Information Exchange offers the kind of comprehensive store and forward mailbox, security, and audit facilities one would expect from an EDI service. Although the ExpEDIte software is provided to users who require it as a part of the service, many forms of input can be accepted, including 2780 and 3780 batch input. IBM works through dealers for the provision of translator packages for PCs but offers its own DataInterchange package for larger systems. Normal access is via SNA or X.25, although asynchronous and synchronous access are also possible. X.400/X.435 will be available from 1993, and Information Exchange also supports the ODETTE File Transfer Protocol (OFTP).

Other services, such as an electronic mail system, Mail Exchange, are offered on the same network. Mail Exchange is not itself an X.400 system, but it can communicate with other electronic mail systems using X.400 protocols. Both the mail system and the EDI system are linked to AT&T Istel's EasyLink service and BT Tymnet.

Subscribers to any of the services on the IBM Information Network buy a package that includes a number of usage units. These units can be used on any service on the network. Charges start at around $1,000 a year for low-volume users.

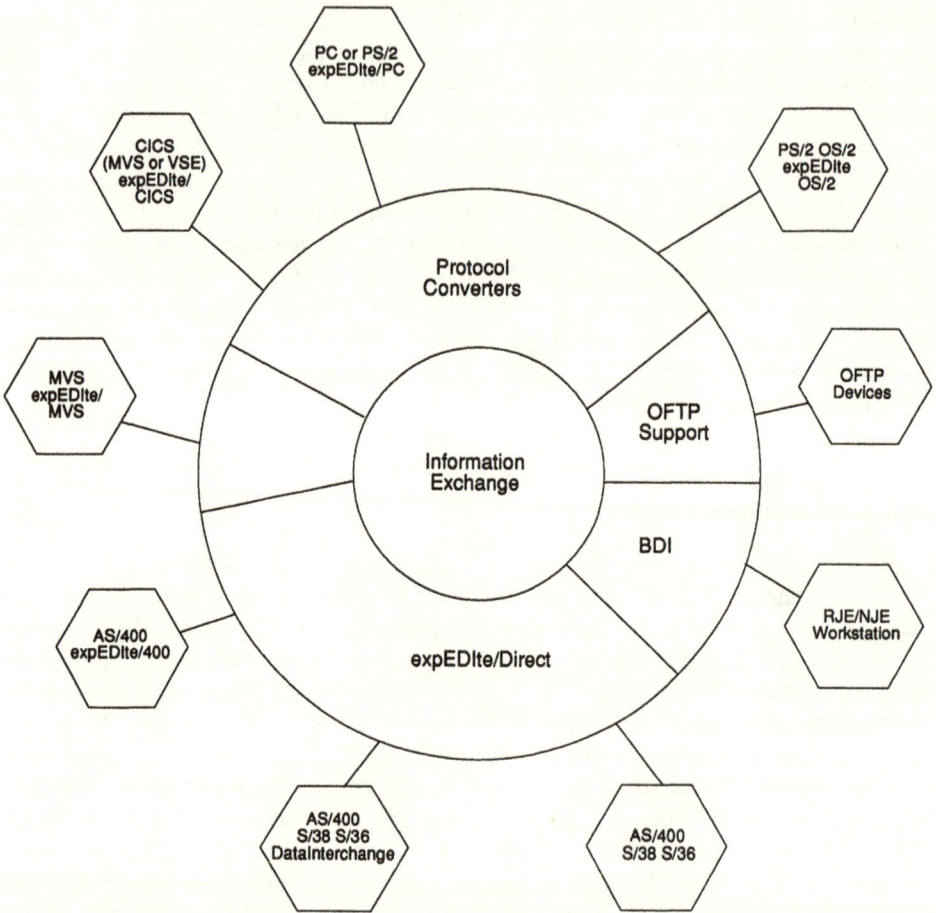

Figure 15.1 IBM Information Exchange.

15.2.5 Summary

Table 15.1 lists the major features of the services we have described.

15.3 EDI SERVICES AND SOFTWARE

In the previous chapters, we have described the scope and nature of most of the major EDI application areas, mentioning where appropriate any special services. Most of these services run either on one of the networks described above or on a private network.

Table 15.1

Major EDI Networks and Services

Network	Year Established	Number of EDI Users (approximate)	Major Participating Industries	Connecting Networks	Costs
AT&T EasyLink	1983	3–5,000	Automobile manufacturing, freight, health	GEIS BT Tymnet, 8 others	Low setup Flat traffic charge
BT Tymnet	1991*	>3,000		GEIS, 20 others in U.S.	No setup; peak and off-peak traffic charges
GEIS/INS	1980† 1985‡	13,500† 4,500‡	Retail, transport	Most	Entry fee interchange, traffic, and monthly charges
IBM-IN	1983	<1,000	Electronics, insurance	Over 20 (mostly U.S.)	Low setup, annual fee, and excess traffic

*In its present form.
†GEIS.
‡INS.

There are also several value-added services aimed at specific sectors that run completely within one of the standard services described above. For example, INS-Tradanet runs an application for transport fleet operators, and another for pharmaceutical suppliers. The insurance services, LIMNET and RINET, described in Chapter 14, both run on IBM Information Exchange.

The EDI services offered by the UK banks to their commercial customers will generally allow access from any of the main networks. Software is provided for PC systems, and the package includes a security system (a smart card or similar mechanism).

Digital Equipment, which has entered the EDI market relatively late, has a joint venture with a number of financial institutions in the UK to provide a service for the exchange of home loan and life insurance applications between brokers and principals. In this case, Digital is making use of its own internal network.

As we have mentioned, when describing each of the main services, most EDI service suppliers also provide some form of software package for the actual communications connection. Users will often also need EDI protocol translation software; this is usually sold as an extra, and costs vary from around $500 to several thousand dollars. One of the most widely used translator packages is Interbridge. It was developed by SD-Scicon for the British Simpler Trade Procedures Board (SITPRO). Interbridge can format and interpret TDI (Tradacoms) and EDIFACT messages. Another general purpose package is ATLAS edi, offered by the French SEMA Group. ATLAS edi will run on common superminicomputers, UNIX platforms, and PCs. It performs data translation and manipulation functions for EDIFACT, TDI, ANSI X.12, and ODETTE, as well as interfacing with the major networks.

15.4 PUBLIC NETWORKS

Although privately operated VANs have up to now taken the lion's share of the EDI network market, the public networks are increasingly able to provide the kind of facilities needed by many EDI users. As networks become increasingly digital, the advantages offered by the VAN operators become smaller.

Public networks have a far wider geographic coverage than any privately operated VAN and reach a much wider range of customers within that coverage. Whereas for some types of operation (such as communications between banks and credit card companies), this is irrelevant, there are many EDI applications where a wider and deeper coverage is very important. In these cases, the extra privacy, service-level guarantees, and value-added services of the VAN must be weighed against the extra connectivity of the public network. Cost will also be an important factor, but there is no simple answer as to which is more economical.

As ISDN becomes more widespread, there is little doubt that it will replace VANs in many EDI applications. The spread of X.400 messaging services could also have a major effect, although the link is perhaps less automatic in this case.

There are also special situations in several countries or areas. In Scandinavia, the X.21 network is a cheaper alternative than a VAN. It operates more closely to real time but does not have the redundancy that will be designed into a good VAN. France Télécom's subsidiary Transpac still has a virtual monopoly on all data communication in France. Transpac's technical systems are extremely good, and the network is completely digital. Transpac itself provides a very efficient EDI service to certain sectors, but the nature of its monopoly means that other services, particularly those involving smaller sectors, take much longer to get off the ground. The retail EDI service Allegro is an example of these (and this is not a small sector): the service was one of the last to start operation of all the major EAN countries, although the Gencod organization that runs it was one of the first to introduce bar-codes in Europe. In Eastern Europe, problems with the conventional wired telephone network mean that greater reliance is placed on wireless (cellular) telephony and on satellite services. The role of the VAN operator is only enhanced in these cases.

15.5 SUPPORT SERVICES

Most of the EDI network and service suppliers provide a full support service, including telephone help and presales consultancy. In several countries, there are also independent

EDI support and advisory centers, run by government agencies, trade associations, or other interested parties. These include:

- Several trade ministry organizations, whose primary interest is in the international trade and customs aspects of EDI;
- The national EDI associations, which operate semi-officially within the EDIFACT framework;
- The article numbering associations, whose interest is in the application of EDI in the retail sector. In Europe, many of these support the EANCOM standards.

To some extent, the various open systems interconnection bodies within the computer industry also promote and advise on EDI, particularly where X.400 connections are concerned. Theirs is likely to be a growing interest.

Chapter 16

The International Environment

16.1 CROSS-BORDER EDI

EDI is international. One of the biggest advantages of the technology is that it allows companies to communicate without regard for borders, time zones, or languages. The key standard in modern EDI—UN/EDIFACT—is the product of an international fusion of standards work carried out on the two sides of the Atlantic. Why then are the vast majority of applications and user groups run within national boundaries? The proportion of EDI messages that cross borders is almost insignificant.

The first answer is that the majority of trade is carried on within national borders. Even companies that are heavily involved in import or export are likely to have more links with companies in their own countries—after all, that is why they are located there. Although European companies are almost ten times as likely to have dealings overseas than their North American counterparts, it is only the smaller countries such as Denmark and Ireland that actually achieve as much business abroad as at home.

One of the largest applications of EDI is in and around the retail business. Retailing is almost without exception a domestic business. The few retailers who have really "gone international"—the petrol retailers or companies such as Sweden's IKEA—generally run their day-to-day operations on a national basis but exchange data periodically from one country headquarters to another.

Importers and exporters are in fact less likely to use EDI than domestic traders, because it is organizationally more difficult to set up systems across long distances. The pace of development varies greatly around the world, and this tends to lead to the continued use of procedures and technologies (such as telex) many years after they cease to be used in domestic trade in any of the countries involved.

There are also practical and technical difficulties: problems obtaining reliable telephone connections, for example, or incompatibility between the communications networks used for EDI in the two countries. The same software may not be available or may not be supported in both countries, and, despite standardization, no one is sure whether two different packages will produce identical message structures.

Traders (and their representative bodies) often underestimate how much business methods differ between countries. Differences in payment terms are easily identified: they typically range from 30 to 180 days. More difficult to judge are questions such as the structure of distribution, the use of free samples or discounting (to whom, how much, and under what conditions). While EDI does not itself dictate any one way of doing business, it takes a very complex system to embody all these factors, which may often require a high level of personal judgement.

There is a tendency for EDI to impose a greater level of structure and hence rigidity on a company's business relationships. This, of course, may mean that the costs of some practices that were previously tolerated are exposed and that managers then decide to eliminate them.

One case where these cultural or structural issues are more likely to be seen as an advantage is for EDI between the plants and offices of a single company or group. Although this is often not regarded as "real" EDI (which should be between trading partners), the plants may each be quite independent profit centers, so that EDI is a significant competitive tool, giving group companies a real commercial advantage in trading with one another. Each national company is likely to have other trading relationships within its own country, thus leading in the long run to a more international network.

The advent of electronic funds transfer and just-in-time production have greatly affected the payment methods used in domestic trade in several countries. In international trade, however, the time scales required for new techniques to be widely accepted are very long: the procedures and terminology used are often seen as both arcane and archaic by those not involved in the business.

The general liberalization of trade that has occurred since the first General Agreement on Trade and Tariffs (GATT) has led to considerable pressure for simplification of these procedures. This was one of the major motives behind the establishment of the TEDIS project in Europe and of several national agencies that promote EDI. Changes in customs procedures have become necessary, particularly in the European Community and the North American Free Trade Area (NAFTA). These should make it very much more attractive to exporting and importing companies to use EDI for communicating with each other and with customs and transport companies.

16.2 INTERNATIONAL COMPARISONS

The opportunity and need for EDI systems varies a great deal between countries. Some of the most important factors are:

- Telecommunications. EDI requires a good telecommunications infrastructure, prefer-ably including some value-added networks and similar service providers. At the very least, the networks available should be able to handle data at 9600 bits/second or above and be highly reliable. A digital (packet switching) network will be necessary for good cost-performance ratios.

- Competitive markets. EDI is to a large extent a competitive tool. It helps to cut costs, reduce delays, and cement the links between customer and supplier. EDI is of no great importance in monopoly or cartelized markets nor where the relationships between customer and supplier are fixed by outside factors.
- Factor costs. This is the economist's term for the relative prices of labor and capital. Where capital is relatively cheap, it is better to spend money on data communications projects such as EDI that save manpower and provide a better service. In countries where manpower is plentiful and capital scarce, it is better to use more people to input and check documents, and to save on equipment.
- Language and social habits. In the end, once all the objective factors have been eliminated, some countries' methods of working depend more on their attitude to life and business than on economic pressures. In some countries (and indeed in some sectors) people simply feel that it is more important to spend time talking to customers about their orders than to deliver the order faster. A maverick entrant to a market (often a foreign competitor) may upset this kind of view, but it also sometimes reimposes itself.

16.2.1 North America

EDI and indeed the whole practice of computer-driven businesses originated in the United States. Scale was an important factor here (there are just more large companies) as was the fact that American businessmen were more willing than their European counterparts to allow their businesses to become less personal. Indeed, this objectivity was often deliberately sought as a goal. Although the two approaches have over time grown much closer together, it is still easier for an American business to be driven by a computer system, and even its customer's computer system, than it would be in Europe.

The telecommunications market in the United States is highly developed and liberalized. Almost any technically feasible service is available, and many of them are very competitive. Prices are much lower in North America than in Europe, with Japan and East Asia in between (cheaper for some services but more expensive for others).

The jobs market is very fast-moving, and good people command high salaries. Capital, on the other hand, is plentiful and, at the beginning of the 1990s at least, remarkably cheap. Distances are enormous, and it is noon on one side of the country before the other side starts work.

All of these factors make the United States and to a lesser degree Canada ideal markets for EDI. Indeed many EDI applications started in these countries. Transportation EDI, in particular for rail and road transport, are widespread and advanced in both countries.

The same scale that makes projects possible in the first instance can, however, make their development a lengthy process. The actual number of EDI users is certainly lower than in Europe and probably not much higher than in the United Kingdom. National and international coordination of projects is a major headache; many industries and industry groupings are quite regionally oriented. Many senior American businessmen have never dealt with companies abroad and have no reason to understand differences in legal or financial systems or in business practices in other countries.

Projects are often sponsored by national agencies, such as the Department of Defense. The need to accommodate all conflicting interests and to demonstrate openness at all stages makes the development of systems and standards a lengthy process, and the systems are often very cumbersome as a result.

With that caveat, the ANSI X.12 standard is a close parallel to UN/EDIFACT, and these two are now virtually the only standards in use for new projects in North America. Usage of EDI on the main networks is growing rapidly, as more EDI software and service vendors form alliances and link their systems into these networks.

16.2.2 Japan and East Asia

The pattern of development for all computer communications in Japan has been quite different from that in North America and Europe. The open-systems principle is only now beginning to be accepted, and proprietary, closed systems have been the rule. These systems often make up for their lack of openness with exceptional hardware performance and security.

Many suppliers in Japan are tied to one major customer or to a group of companies; this customer often owns part of the company and may have appointed one of its former executives as a director. Japanese business is dominated by these large trading groups, often centered around a bank or large manufacturer. It is therefore no surprise that EDI in Japan is dominated by closed user groups, often run by a major company to cover all its relationships with its suppliers. Over 500 manufacturers have in place electronic ordering systems (online networks for order placement), and there is a similar number of local VANs covering several related sectors within a local area.

The use of UN/EDIFACT and other international standards was approved by the Ministry of International Trade and Industry only in 1991, and special hardware is being developed to provide integrated EDI message handling within an ISDN environment.

Elsewhere in East Asia, EDI has not been widely used. Although these are supremely competitive markets, some coordination is required to develop standards and promote an EDI system. Singapore has the nucleus of a national EDI system, but it looks as though Hong Kong may become the first territory in the world to establish a single integrated EDI system covering all sectors and applications. The first stage, which will be implemented during 1993, will give users cheap EDI access to customs and government trade bodies.

16.2.3 Europe

The European countries are not a homogeneous bunch. They vary from the highly industrialized countries that form the core of the EC to the still quite rural economies on the fringes. In most of Western Europe, the telecommunications service is very efficient and a wide variety of services is available, despite the fact that, in many countries, telecommunications is still a state-owned monopoly. Costs, although higher than in North America, are highly affordable to most users and falling steadily. In Eastern Europe (and the eastern half of Germany), development of the telecommunications infrastructure is a high priority.

Relative factor costs have varied more than normal during 1992, as interest rates and labor market measures have been used for political and fiscal purposes unrelated to their value. This makes it difficult to pass judgement on the long-term effect of the current low interest rates, which would normally make it exceptionally attractive to finance investment in equipment.

Attitudes to computer communications are changing rapidly. At first, many companies viewed such changes to traditional ways of working as a threat. The apparent success of those who were the first to use them caused the change, but it was then found that copying their methods did not yield the same advantages. Companies now realize that customer service and product differentiation are the key issues. The first companies were successful precisely because they offered something different.

The range of diversity in attitudes, languages, and financial situations in Europe is extreme. This is where the greatest potential advantage of EDI lies. Nevertheless, many of the most successful projects have exploited one large sector (such as the grocery sector in the United Kingdom), and then only when penetration in that sector passed a critical mass did it expand into other sectors.

The UN Economic Commission for Europe and the European Community's TEDIS program have provided a focus for international development of EDI. Several specific projects, in particular those relating to customs cooperation and the simplification of trade procedures, have been developed under the aegis of UN/ECE and with funds provided partly by TEDIS.

16.3 INTERNATIONAL SYSTEMS AND NETWORKS

Most of the main EDI networks described in Chapter 15 offer a range of international connections including the US, Canada, and major European countries. Some range wider than this to include Australia and East Asia, for example. But EDI support is likely to be limited. Those networks that are able to offer a full international service have an advantage over separate networks connected by gateways, in that they are able to offer a more integrated audit, security, and billing system.

New satellite-based services may come in as VSAT (very small aperture technology) terminals become more common. In Europe, this will require much more liberalization of the satellite data communications sector. Satellite beams do not understand the limitations of national borders, although the beam shapes of some of the newest generation of satellites are tailored closely to the coverage required on the ground. U.S. satellite service operator Microspace is seeking to offer a data broadcast facility to sites in Europe, uplinked from a site on the other side of the Atlantic.

The main EDI services offering international coverage today are the customs networks (although each trading company would normally expect to deal only with its national customs office) and the international purchasing networks of very large EDI users such as IBM. Other systems, as we said at the beginning of this chapter, are normally operated within national borders, although they may exchange data with systems in other countries where necessary. The major advance that has been made in EDI standards in the past three or four years is that those international exchanges are not only possible but should not really be any more difficult than their domestic equivalents.

Part III

Chapter 17

Opportunities and Problems Revisited

17.1 BENEFITS, PROBLEMS, AND ISSUES

At the beginning of this book, we reviewed the benefits that could be gained from introducing EDI. Now that we have looked at the technology and the way it is used, it is time to take stock of the progress that has been made in resolving the problems and taking advantage of the opportunities.

In Chapter 2, we list some of the main direct benefits that could be gained by introducing EDI into almost any business. These included:

- Faster transaction turn-around;
- Less paperwork;
- Savings in staff time; and
- Fewer errors.

There were also potential problems:

- Legal issues (Is an EDI order a legal contract?);
- Security (How can we make an EDI system secure?); and
- Standards (which can be time-consuming to prepare and may not fit everyone's needs exactly).

Some of the most important issues, however, were cases where one party might gain while its trading partner loses or gains much less. These are benefits or problems depending on the viewpoint of the observer. They include:

- Reductions in inventory that may simply shift the burden of carrying inventory to a supplier;
- Suppliers can be paid faster with EDI, but this means that the customer has to pay faster; and

- Closer links between customers and suppliers, where one of these is more powerful, however, this may have the effect of locking the smaller party into a relationship in which his bargaining power is reduced.

We will now look in greater detail at how these benefits, problems, and issues have worked out in practice, and how we can expect them to develop in the coming years.

17.2 COMPETITIVE OPPORTUNITIES

The first companies who introduce EDI in a given sector can gain a competitive advantage from the cost savings that the technique can give. Before long, however, that company's competitors will also bring in EDI, and the competitive advantage disappears.

In fact, the pioneers will carry more of the burden of setting standards and convincing customers that this type of operation is beneficial to them. Paradoxically, they may be less well placed to extract the greatest competitive advantage; it is rare for a company that is able to lead its sector in a technical initiative also to be the strongest in marketing initiatives. EDI affects all areas of the business process. This area is not sufficiently well studied and there are few firms who even consider it in their decision-making.

Companies who have had EDI imposed on them as a condition by their customers or dominant suppliers may not even recognize the short-term advantage.

As EDI becomes widespread within a sector, it becomes not an advantage but a necessity for operating in that business area. Companies that do not use EDI suffer a disadvantage, unless they can find other ways of making similar savings or of differentiating themselves in a positive way to their customers.

Marketing in the 1990s is very much concerned with differentiation: the idea that a company's products or services must be distinctive in some way. One company may be the cheapest, another may offer exceptional performance and a third the best service. There will be a place for all of these three in the market, and all can make reasonable returns by emphasizing their strengths. If, however, they all try to compete only on price or on performance, this leads to a war that erodes profit margins and satisfies some customers but not others.

In fact, there can come a point where EDI is so much a common factor in the business that companies find it hard to drive administration costs down further. It can also make it difficult for a manufacturer, for example, to differentiate himself on quality of service when everyone's service is offered through the same computer system. A company wishing to differentiate itself on service would have to distance itself from common industry moves like this.

17.3 PERCEIVED PROBLEMS

The first problem that potential users see in EDI is technical: What hardware and software are we going to have to buy? Are there standards covering our line of business and the way we operate? Will anyone in the company understand what we have to do or will we have to bring in a consultant? There are relatively straightforward answers to all of these questions, and we hope that most of them are contained within this book. Whereas it will obviously be necessary to investigate individual packages and to seek quotations from service suppliers, these are commercial factors that should not divert anyone from a decision to pursue EDI.

A key point to bear in mind is that EDI is not a technology. It is a concept that makes use of established technologies in computers and telecommunications. The key to EDI is the use of a pre-agreed structure and vocabulary for the messages: the EDI standard.

Although there are literally hundreds of committees and subcommittees working around the world on EDI standards, only a limited number of messages has actually been finally agreed. It is sometimes argued that a critical mass of messages has not yet been reached. In practice, standards are simply not a problem to the vast majority of commercial EDI users: there are standards for nearly all the things they are likely to want to do, and, even if the standards are often rather complex, this means that they are able to take into account most ways of doing business legally in the western world.

Many areas within the standards can safely be ignored and many fields are optional, so that users whose needs are simple can quickly extract what is necessary to meet their requirements. The translation packages that are usually employed to convert users' data into the correct EDI format take account of all these factors.

The same is not yet true for technical EDI, which has the potential to be at least as important as commercial EDI in the long term. Here the critical mass has not been reached, both as far as messages and users is concerned.

The legal status of an EDI exchange is still inadequately defined; although various bodies have ruled on specific questions, an EDI message is still generally inferior to a paper message in the eyes of the law.

Nevertheless, many companies trade regularly and securely through EDI. Provided that good current practice is followed by all parties, problems of a legal nature are unlikely to occur, and these can in nearly all cases be resolved by arbitration. The UNCID rules define in general terms what is meant by good current practice, although the specific details have to be looked at in each case.

Most EDI services have a form of contract that attempts to define the responsibilities of the parties, and many independent EDI users are now using interchange agreements, as discussed in Chapter 8. These will nearly always provide for arbitration in disputes, thereby avoiding the case reaching the courts, where the admissibility of evidence becomes more important than the merits of the case.

The security of EDI systems is very important, especially where financial transactions are concerned. Computer systems can be made as secure as necessary—considerably more secure than any paper-based system. Most EDI systems (even many financial systems) use a rather low level of security, but are felt adequate to protect against the main risks of access to the system by unauthorized users. This is often complemented by a comprehensive audit system, so that disputes or breaches of security can be traced and followed up. We believe, however, that customers will start to demand more security on all types of computer networks.

Smart card-based security systems are now becoming popular; they are inexpensive and remarkably unintrusive for the user, and, unless something better appears, we expect that these will become a normal feature of EDI systems, even for routine operations. Smart cards are best at providing security in interactive operations, but can also be used to provide functions such as message and terminal authentication in a higher-powered real-time transmission security system.

There are now few if any social or attitudinal barriers to EDI. In order to consider EDI, businesses must already use computers. While the integration of the EDI system with existing procedures and organization must be, as we will discuss further, carefully handled, EDI is no worse in this respect than other computer-based systems.

One residual problem is that of the inequitable gains arising between the two trading partners if EDI is used for accelerating the payment process. EDI, however, cannot be blamed for this, and a decision to make payments earlier or later is independent of a decision to reduce the paperwork and processing time associated with that payment. While it may remove ninety of the hundred most commonly used excuses for not making a payment, an EDI system can still be designed to give the payer complete control over the timing of the payment.

17.4 UNDERLYING PROBLEMS

Despite this healthy balance in favor of EDI, it is wise to be aware of the underlying problems that can exist.

17.4.1 Loss of Differentiation

Earlier in this chapter, we touched on the way EDI can make it more difficult for a company to differentiate itself from its competitors in terms of service. This problem can become even worse, however, if EDI is allowed to replace some of the more traditional selling tools. Screen-based quotations, which are already used in sectors such as insurance, travel, and financial services, completely eliminate differences between products that are not specifically covered by the search parameters.

Even such simple marketing advantages as a good product name lose their edge if the customer does not have to remember the name. There is a strong tendency for all products to compete on price. As we mentioned earlier, this is not generally healthy for the sector and does not necessarily meet customers' requirements any better.

17.4.2 Design Around People

Great care needs to be taken when designing EDI systems in existing operations. The business is still run by the people and not by the EDI system. It is often wise to design the system so that initially it is doing exactly what the manual system does. This is often wasteful, but makes it easier to understand, and the two systems can easily run in parallel. Modifications that may have been envisaged or even designed from the outset can be introduced later with fewer problems. In other cases, it is probably better to use the introduction of EDI as an opportunity to consult with existing staff in detail and to use their experience during the design process. It is easy to implement changes to procedures that the staff have already acknowledged to be necessary or desirable. They will often explain the changes to their counterparts at the trading partner, and further feedback may come from that source.

The problem is often that, like the road to Tipperary, it would be better not to start from here. Existing systems may have built into them hundreds of inefficiencies caused by the limitations of older communications methods. It is very rare, however, for a company to have the luxury of designing its systems from scratch, without having to take into account the systems of its trading partners. Even new companies rarely implement EDI and similar systems initially, if only because they can handle small volumes of business manually without difficulty.

17.4.3 Future-Proofing

When specifying or installing any computer-based system, it is important to ask the question: What will we be doing in five years time? How much will the market have changed? This is of course a business decision unrelated to EDI. But EDI should not constrain the business from moving into new areas. Computer technology is advancing as fast as any sector, and data communications is one of the fastest-moving sectors within computing. So it is unlikely that data communications hardware or software will be a limitation.

Application standards such as EDI cannot move as fast as the underlying technology. So here there is a need for a little anticipation; pioneers in any field may have to wait a while. But pioneers by instinct are unlikely to be put off by a shortage of specific standards, particularly when the general direction and basis of the standards is soundly established.

Some of the more radical suggestions in Chapter 18 may require a more fundamental rethink, but these would be beyond the immediate lifetime of equipment and software being purchased today.

17.4.4 Connectivity

There is an important choice to be made today between a value-added network, which provides extra facilities and functions to a limited number of connections, and the public network, which has the widest possible range of connections.

Most EDI systems today have opted for the VAN approach. The VAN operators are making strenuous efforts to extend the range of their services, either by installing further nodes or by linking into other networks. In the latter case, problems of compatibility can arise, often only showing up when some unsuspecting customer tries to perform an operation that functions differently on the two networks.

The functions available from the public networks are being extended, in the largest countries at least, and their reliability and performance for data applications is also increasing. EDI applications rarely demand very high performance, and so this route may be perfectly adequate for many systems. It is already the most common route for EDI connection in Scandinavia.

17.4.5 Business Requirements Must Stay in Control

EDI acts as a useful discipline for many companies. On one hand, it demands that the rules of the business and of its relationships with its customers are set down and makes it difficult (it should not make it impossible) to override these rules.

On the other hand, it is important to ensure that the rules are set by the company's business requirements and not by the EDI system. It does sometimes happen that a computer department is given the job of interfacing existing systems with an external EDI service. They will do this in the easiest way—it is not usually difficult if translation and EDI software are brought in—whereas if a full systems analysis were performed it would show many little quirks, controls, and overrides in the existing system that must also be reflected in the EDI system.

In Chapter 8, we emphasized the need to review the business systems before final specifications for the EDI system are drawn up. Users must also be prepared to change the EDI system if the environment changes or if problems are found. EDI must never be allowed to become an end in itself.

17.5 BUSINESS OPPORTUNITIES

The real opportunity afforded by EDI is this ability to rethink the way the business is carried on, freed from many of the old constraints. A really fresh approach is needed so that the business can concentrate on essentials, without taking into account practices that have only come in because of the slowness of existing systems.

Approval of payments is one example of this. Existing systems typically require a minimum of 30 days to make a payment, because the invoice must first be received through the post, then entered into a ledger, then allocated a cost code and sent for approval. The manager responsible for approving it may be away for a few days, so that it is two weeks from the invoice date before it reaches the accounts department again. Then the whole process has to be repeated with the check that is used for payment. Because of this, 30 days has become regarded as the minimum "normal" payment period. For some large companies, even this is difficult to achieve. This is the type of assumption that can be removed by a full EDI design.

Similarly, where a company is asked for a quotation or delivery estimate, they will usually add some contingency to the figures just in case things change between the quotation date and the order date—for example, another customer order coming in or the cost of goods rising. With EDI, quotations can have a lifetime of less than 24 hours, so that they can be more accurate and less contingency is required. If the EDI system is linked to the production control system, planning can start as soon as the estimate is issued.

With EDI linked to a production control system, it is much easier for a manufacturer to produce a variety of products and to change production schedules dynamically in line with demand. Customers appreciate this added flexibility, and it also means that expensive production equipment can be used more intensively and effectively.

This ability to plan production in a flexible and varied way is only useful, of course, if the equipment is capable of the full range of tasks envisaged. Machine tools have become infinitely more flexible in the last twenty years, performing different functions with a five-minute tool change and program download. Injection molding machines have always been good at producing different items within a size range, but here the progress is in the ability of tool-makers to design tools that can be changed quickly and that will not need long and precise set up. Filling machines are relatively inflexible, and production runs still are quite long in this sector.

Outside manufacturing, it is much easier to exploit the extra flexibility afforded by EDI. Service industries are particularly well placed.

Whereas we said earlier that a potential problem with EDI lay in the loss of opportunity for differentiation by the normal means of quality and speed of service, the EDI system itself can be the vehicle for traders to provide new services to their customers, and hence to differentiate themselves in new ways.

Again, this is a question of appreciating the business process requirements—here, the customer's business process. EDI offers a real opportunity to become closer to one's customers and, in the long term, to develop new forms of business relationship.

This is already happening extensively in the retail and transport sectors. Several supermarkets have exclusive arrangements with fresh food suppliers, under which they share point-of-sale data with them in return for guaranteed deliveries. Transport companies give their customers software that will send them details of a shipment as soon as they are available.

17.6 A NEW VIEW OF EDI

What is developing is a new view of EDI, as depicted in Figure 17.1. We have gone from the purely technical, cost-saving view of EDI to using it to obtain a competitive edge. As that competitive edge is eroded, traders are seeking to use EDI to build new forms of relationship with their suppliers and customers. This in turn allows many of the old rules of the business to be turned on their heads, so that the business can concentrate more on essentials and less on the mechanisms.

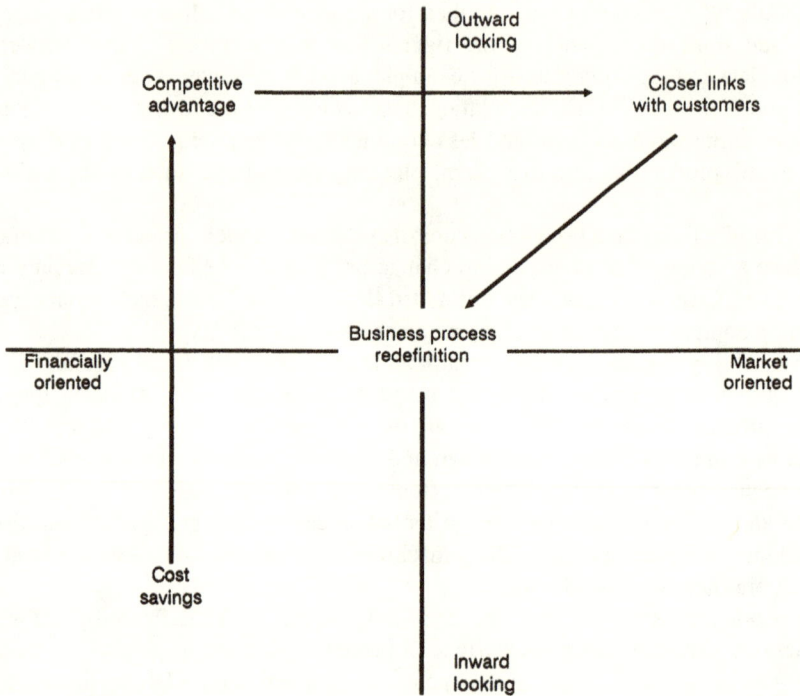

Figure 17.1 A maturing view of EDI.

The first phase is very inward-oriented and concentrates on the financial advantage: we can save money through using EDI. Any other advantages are seen as ''soft'' and are not taken into account in the financial justification.

Once the marketing department becomes aware of the possibilities of EDI, these are widely publicized and start to be used to promote the business. Although the emphasis is still on the hard-nosed cash advantages, be it reduced inventory or faster order processing, this phase is marketing-driven and looks outward, to the effects of our EDI operation on our competitiveness.

It is usually only after embarking on this phase that we realize that EDI is an advantage to our trading partners as well, and that they could also benefit from our experience. We start to point out the ways in that we could exchange data to our mutual advantage, and often at this stage there will be visits to each other's premises and technical discussions between software personnel. Building such new relationships with our customers is outward-looking and market-driven.

The most mature view of EDI, to date at least, is to consider it as allowing new ways for the business to operate. This view looks as much inwards as outwards, and takes into account the financial, marketing, and other (e.g., production) aspects of the business. EDI becomes a part of the strategic framework of the business.

Chapter 18
The Future of EDI

18.1 EXTENDING THE USER BASE

In the short term, EDI will continue to develop along what has become an established path: within existing sectors, the ''pyramid selling'' effect will continue to generate new customers and extend the user base.

At the same time, seeds will be sown in new sectors. There will be a period of one or two years before suitable messages are devised, adapted, or selected, but during which large potential users will be trying out the technology and recruiting partners into the scheme, so that when it becomes operational there is a critical mass. From that point, the pyramid can take over.

The analogy suggests that EDI service providers do not have to market or sell their wares: existing users do this for them. This is true in some cases, but more often there is a shared responsibility. Systems promoted by industry associations, for example, are often marketed by those associations to their members. It is very unlikely that an EDI bureau would take on this type of activity; however, they would be expected to provide extensive backup to any promotional activities carried out by the trade association.

The EDI network companies are more active in selling other network services to existing EDI users, or conversely, selling EDI to electronic mail users.

As the market develops, however, this will surely change. EDI will become a commodity, particularly as networks are interconnected and as multiple services become available within single sectors. Service and network providers will have to compete for new customers and for new business from existing customers.

For the moment, however, the market is so far from saturation that these considerations are still some way off.

18.2 NEW APPLICATIONS

The future of EDI depends on new applications being developed. In the immediate future, the most likely candidates are technical EDI applications, particularly in companies already using EDI for commercial applications.

The computer and electronics industries, as we mentioned in Chapter 12, are increasingly keen users of EDI today, often in closed user groups but also now participating in more general, EDIFACT-based systems. EDI standards covering the transfer of electronic design data and specifications are now becoming relatively usable, and this could become an important area in an industry that uses many subcontractors.

With EDI, the design process can allow interaction between different teams working on the same project—an increasingly common way of working on larger projects.

EDI is likely to be used increasingly for the distribution of manuals and updates, as in the CALS project. Software updating is a major problem for many developers, but its impact could be greatly reduced through the use of broadcast EDI systems.

Mechanical engineering data transfer standards have further to go, and it is less clear that there is a substantial user base that would be willing and able to make use of them. Even with government support for projects such as CALS, this application is likely to be a slow starter.

Other applications could arise from the freeing of boundaries that EDI permits. For example, it could lead to an increase in international retailing, which is so far not very widespread. With better data communications, however, national boundaries would be less important.

There could be scope for major new systems just by exchanging applications between sectors and countries. This is the kind of development that the international EDI networks could undertake, since they often have both sets of customers on their networks.

Government departments and other public bodies would be able to provide a better service and reduce their own costs at the same time if much of their routine work were carried out by EDI. Although governments were among the first large-scale computer users, their systems have not developed as fast as those in the private sector, and they are very often still completely batch-oriented. Even with this limitation, much government paperwork could be saved if data and queries could be submitted electronically.

18.3 TECHNOLOGICAL DEVELOPMENTS

18.3.1 Computing

Recent advances in computing power will continue. Although the spotlight has been on the increased power available at the lower end of the computer market, particularly in PCs, much progress has also been made at the higher end, with even supercomputers becoming much more powerful. It is the changes at the lower end, however, that have made computing more accessible to a wider market and thereby expanded that market many times.

EDI itself is relatively undemanding in computer power. Computers are good at moving data around, and do this without much effort. Power is needed for such tasks as graphics handling and making sense of abstract concepts. Special graphics processors are now being used to handle the first, while neural networks are being designed to help with the second.

Neural networks are supposed to act like the brain: they set many processors to work on different aspects of the problem at the same time to see how many can come up with a solution, or help each other in the process of building a more complete picture. It is intriguing to ask how EDI might interact with a neural network, or whether you could actually have EDI as a part of such a network.

Partly as a consequence of the development of smaller, more powerful machines and the software to drive them, more computer systems are becoming distributed—that is, the intelligence is put where it is required and not in some remote computer room. This is very important for EDI, since distributed systems need to use more open standards. EDI as a concept covers all the layers of an open systems model, and could therefore be used within a system as well as for communicating with the outside world.

Smart cards are the logical extension of the distributed-computing concept. They are already used for security applications in EDI, and we expect that they will be used for many more personalization tasks. A complete protocol could be contained on a single smart card.

The moves towards open systems for all types of software applications are only helpful for EDI. As awareness of EDI increases, software designers will take it into account, in the way that they are now taking into account multi-user requirements. Implementation of EDI is made much easier if suitable "hooks" are left in the software for the addition of tasks later.

18.3.2 Data Communications

Technical advances in telecommunications have slowed down in the last few years. This is partly because the technology had pushed ahead of its commercial applications; much more money was being made from less advanced technology than from the most up-to-date. It is now the turn of the commercial side to make progress.

Analog, copper-based systems have reached the limit of their development, both technically and commercially. Future hybrid devices could, of course, reverse this again. For the moment, though, the future lies with digital communications, and increasingly with optical fibers and radio-based systems rather than with copper cables.

The technology is available to provide data transmission anywhere at millions of bits per second—a thousand times faster than most of today's circuits—but the rate at which this technology can spread is limited by the need to grow a complete new set of industries around it, and by a huge established telecommunications industry that depends on maintaining a price differential in order to survive.

Radio communications, using VHF/UHF terrestrial systems or satellites, are growing faster than fixed links. This reflects not only a different regulatory system, but also the growth of a more mobile society.

The big question surrounding the future of telecommunications for EDI is the extent to which public networks will play a role, or whether private VANs will continue to dominate.

With increasing deregulation of the telecommunications sector, this could become an irrelevant question. The "public" networks, which may often be run by private companies, are offering increasing levels of value-added service, and are becoming much more aware of the needs of data transmission. The "private" VANs are expanding by linking together and giving easier access through the public and other networks. The two cross over at the point where VAN operators are operating specialist network services on behalf of the public operators.

In the medium term, it is likely that a fully international electronic mail service (using the X.400 standard or some derivative of it) will develop, initially on private VANs but then gradually on the public services. There are already partial systems in existence, but for a full international system to develop, further work needs to be done on removing the differences between national implementations of the X.25 and X.400 standards.

As it is relatively easy to add store and forward facilities to any packet switching network, the next step would be a public EDI service. This could be implemented with very little technical difficulty; again, the problems are more likely to lie in the commercial implications of upsetting the established order.

We believe, though, that the long-term future of data communications systems lies with a "layer" concept, as shown in Figure 18.1. We start with a very general network that simply offers physical connection into the global system. This is followed by a level dedicated to addressing and directory services; customers may choose a simple and direct system, such as today's international telephone numbering system, or more intelligent systems tailored to their specific needs.

Above this level we start to distinguish between different types and rates of traffic: from voice to voice and image, transaction-oriented data, and bulk data. Then we have levels that acknowledge the type of data: payments data may be separated from retailers' price lists or orders. Within each type of data there may be further hierarchies: for example, merchant-to-bank, interbank, and finally interclearing system. Each of these would have its own availability and security requirements, each progressively higher in the case we have mentioned.

Several of these layers or subsystems could use the techniques we today know as EDI. In practice, however, EDI standards will have to be modified to meet the requirements of other layers and other systems developments.

18.3.3 EDI Software

The current trend in EDI software is to take account of the need for interactive EDI. This implies "event driven" software, that can dynamically change priorities to handle

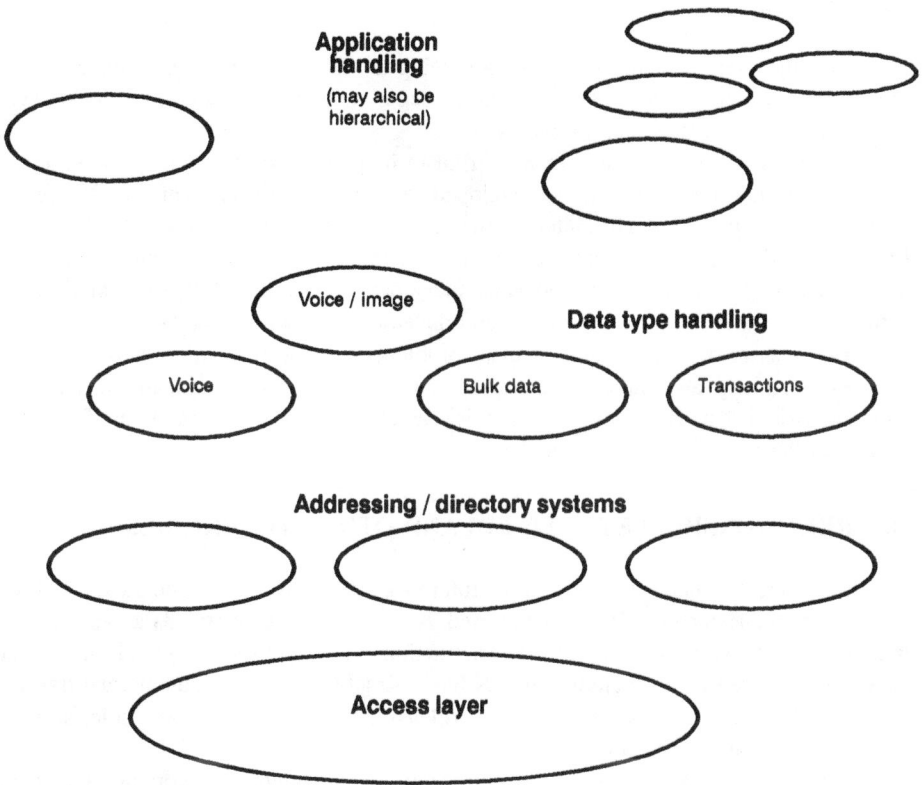

Application handling

(may also be hierarchical)

Data type handling

Voice / image

Voice

Bulk data

Transactions

Addressing / directory systems

Access layer

Figure 18.1 Layers of data communications.

interactive tasks more effectively. Although it is in principle a good idea to have control of priorities, the event-driven facility is not required for the vast majority of EDI, and may indeed make it less efficient. Both need to be available.

A high proportion of today's business computer applications are database-oriented, and in many of them there are far more enquiries than transactions. EDI has so far concentrated on transactions that change the database, rather than on enquiries. We believe that a growth area for EDI will take the form of a structured interchange of queries, using protocols derived from a standardized structured query language.

This would enable a complete new set of applications, based on the principle of the industry-wide database. Such a database could cover product data, price and delivery quotations, and contractual and servicing details. We will come back to this later when talking about new ways of doing business.

18.3.4 Standards

New messages and vocabulary for EDI are being developed at an accelerating rate. The process for submitting and agreeing on messages and directory entries is now well-established, and operates internationally.

A key area for the development of EDI in its present form is clearly the ease and speed with which messages can be developed for new applications. This is often felt to be one of the drawbacks of a full international standardization effort such as UN/EDIFACT. EDIFACT's defenders will point out that the process of international standardization is necessarily long, that messages and vocabulary can be used at all statuses, and that in extreme cases users can use the syntax and develop their own messages.

One way to overcome some of these problems may be to use a "super-syntax" or parameterized language that could permit the language to grow dynamically, as a natural language would. This is not, however, a quick development, nor one that would fall easily to the developers of the current standards.

18.4 RELATIONSHIP BETWEEN EDI AND OTHER TECHNIQUES

The most important change we see in the future for EDI is that the boundaries of today's carefully defined concept will become blurred. As EDI merges with bulk data transmission, interactive systems, and other real-time transaction-oriented systems, the definitions that have been adopted for it, including words like "structured data," "commercial transactions," and "store and forward" become irrelevant. EDI is a part of the whole fabric of intercomputer communication.

The striking success of EDI has been to set down elements of a common language, to define the way that language can be used for specific transactions, and the media or transport mechanisms that can be used for conveying the message.

EDI as a concept is also a part of the open systems interconnection idea. EDI is easier to grasp than OSI, however, and unlike OSI it can be implemented by users rather than by manufacturers. EDI may turn out to be the vehicle by which OSI is converted from theory into practice.

18.5 NEW WAYS OF DOING BUSINESS

You will have realized from Chapter 17, though, that the most important contribution EDI can make is often to change the assumptions underlying many business practices, and allow a company to concentrate on essentials rather than on the received wisdom about an industry. These better-business practices include:

- Greater accuracy;
- Prompt service and knowledge of the full situation;
- Use of internationally accepted standards;

- More openness with customers and suppliers, leading to closer cooperation;
- The opportunity to review and optimize the business process; and
- The opportunity for a better-educated staff, and for them to have a wider view of the business.

Where interchange agreements are used, particularly internationally or across sectors, the status of all forms of computer data are clarified, in practice if not in theory. One major advantage is the availability of immediate acknowledgement for orders, deliveries, invoices, and payments, thus clarifying a possible area of doubt or dispute.

From a strategic point of view, EDI helps to link "islands of automation" into a complete system. Although this requires additional software in other areas, EDI standards help to define the interfaces and, to some extent, the software structures required.

One dramatic change that should be possible only with the help of EDI is to alter the way international trade operates. At present, this sector suffers from several business practices that were born of necessity decades, or even centuries, ago. With the common language of EDI, two traders, their banks, and the customs authorities can communicate in less time than it takes the goods to travel from one border to the next. This could potentially remove the need for freight forwarders altogether, particularly if transport companies also moved comprehensively to a screen-based enquiry, quotation, and booking system.

What would in fact happen if markets such as this moved to fully online systems is unclear. If it really did lead to an approximation of perfect information, then the economists' dream would be realized and the market would work better, to the advantage of supplier and customer alike. It would certainly remove some forms of distortion and would lead to more price competition.

As we have indicated in previous chapters, this is in principle a good thing, but not all the effects are beneficial. If all competition is centered on one factor, such as price, then the profitability of the sector as a whole is weakened. EDI users must find ways of using it to increase differentiation—by providing additional services through the EDI network, for example—rather than accepting the loss of differentiation.

This is frustrating for standards developers, for whom any attempt to use the standards in a different way is a distraction. But it is essential to a really mature view of EDI and its long-term future.

The future of EDI lies in its ability to promote, not just accept, new forms of business relationship and new ways of doing business.

Chapter 19
Review and Summary

This chapter is a summary of the main findings of each chapter in the book, and finishes with a checklist of the areas of business that are affected by EDI.

19.1 WHAT IS EDI?

Electronic data interchange is a simple concept that has far-reaching implications covering the technical, financial, marketing, and organizational disciplines within an organization. It involves using direct links between computers to send data, such as commercial documents, design data, and payment instructions that would otherwise today be sent in printed form. The applications cover all types of companies and industries, from banking and finance to retail shops, transport companies, and small manufacturers, to customs and other government agencies. Although many of the most important applications are in trade (buying and selling goods), EDI is also used for exchanging other forms of data between companies, such as electronic designs, marked-up contract documents, or customs returns.

19.2 ADVANTAGES AND DISADVANTAGES OF EDI

EDI saves time and manpower by avoiding the need to re-input the data at several stages. This also eliminates errors that rekeying introduces; errors are particularly common in cases where the operators have little knowledge of the products they are dealing with. The data arrives much faster than it could by post, and acknowledgement is automatic.

EDI is a discipline; EDI operations must be carried out in a certain way. This can be both an advantage and a disadvantage. It increases a company's dependence on computers and telephone lines, and these are not always as secure as we would like.

Some companies have already reduced the delays in their systems by using the fax, which may be cheaper to operate than paying the costs of the necessary hardware and software and the subscription fees to an EDI service. But this does not give the same scope for changing the relationship the company has with its customers or suppliers, and for redefining the company's systems to eliminate unnecessary operations.

While some of the benefits of EDI, such as the reduction in paperwork, are an advantage to both partners in a trading operation, others are seen as one-sided—they are an advantage to one and a disadvantage to the other. For example, if a retailer reduces his inventory thanks to a new EDI system, this may simply mean that the manufacturer has to hold a larger inventory. If one side is paid earlier, the other must pay earlier.

Unless the mutual advantages are compelling, EDI is often imposed by one side: for example, a dominant retailer, oil companies, or paint manufacturers. The dominant partner may well impose it on all its suppliers or trading partners, so that any competitive advantage is short-lived.

EDI can then become almost essential for doing business competitively in that area; there are several sectors where the main customers have actually made it a precondition. Suppliers can subsequently find it difficult to differentiate themselves from their competitors. They must look for ways of using the EDI network to increase rather than decrease that differentiation.

19.3 LEGAL STATUS

As well as any commercial disadvantages there may be, there is a potential problem with the legal status of EDI transactions. The main issues surround the timing of a transaction and the admissibility of evidence. At present, EDI messages, even with the benefit of a full audit trail, have a lower status in law than paper transactions.

In practice, the problems of comprehension that would surround any point of law in an EDI transaction would make any court case expensive and its outcome would be unlikely to bear heavily on the merits of the case. EDI users are therefore advised to specify arbitration in any contracts.

While users are often concerned with the security of EDI transactions and systems, there are adequate safeguards available if required.

19.4 FUNCTIONS OF EDI

The functions of an EDI system are to:
- Extract the relevant data from a computer process;
- Format the data according to the EDI protocol;
- Connect to the EDI service (often a third party);
- Send the messages, which are then left in a "mailbox" for the other computer;
- Allow the other computer to fetch these messages; and
- Decode the messages and input the data to the next stage in the process.

Included in this are communications, buffering, format conversion, security, audit, and accounting and analysis functions. An EDI system must also be highly reliable, have wide geographical coverage, and be able to connect a wide variety of equipment types.

19.5 SOFTWARE

Good EDI software, whether it be a general message handler, a package specifically designed for the network in question, or an EDI gateway tailored to one company's requirements, should avoid the need for users to be involved at the detail level.

In particular, it should not be necessary for users to become familiar with the structure of standards such as open systems interconnection, even though almost all new EDI software will conform to these standards.

EDI software will perform the functions of format conversion, message management, communications (which may need to be specifically designed for the network being used), and security checking. Although password controls and encryption can be handled by the software directly, good physical security is still essential.

19.6 COMMUNICATIONS

Individual EDI users are often not free to choose what communications network they use. Where they can do so, they are faced with a bewildering array, from public telephone networks to leased lines, packet switched networks, and ISDN. Private VAN operators have the lion's share of EDI business in most countries, and they are also able to provide international connections. Their services are more expensive than the public networks, but the additional services they can offer make them a more attractive choice today.

19.7 HARDWARE

EDI systems should be independent of any particular hardware: user systems can be, and often are, simply PCs with a modem, while a host system is likely to consist of a number of mid-range or supermini-computers. EDI hardware can be very simple, and there are rarely any very stringent performance requirements.

19.8 STANDARDS

EDI depends on agreement between users on the language they will use, including both its syntax and vocabulary, the way that language will be used (what messages will be exchanged, what are the contractual arrangements), and on the media (such as a network) that will be used to transport the messages.

In recent years, there has been a massive international standardization effort in EDI. Although many standards do exist and will continue to be used for several years, the dominant standard for international use is the UN/EDIFACT standard promoted by the UN Economic Commission for Europe. This corresponds fairly closely with the American National Standard X.12, and there is a program of cooperation and *rapprochement* between the two.

EDIFACT covers many applications within the trading and commercial environments. Although its syntax can be used in other areas as well, it is less commonly used in the applications known as "technical" EDI—this term covers everything that is not related to buying and selling goods and services.

The standards in these areas are less well developed, although progress is being made, particularly in areas such as computer-aided design for electronic and computer equipment.

19.9 IMPLEMENTING EDI

A company starting to consider EDI must expect a fairly long time scale: we suggest a minimum of six months. The implementation process must be planned, and should include several iterations at each stage in order to produce the best results. Support from senior management is necessary for a successful implementation.

The process starts by determining the terms of reference and forming a project team. They should write a requirements specification and seek budget approval. Once this is granted, they can proceed to a functional specification and consider an interchange agreement with the main trading partners affected.

They will then select a network, and any hardware or software that must be purchased. Training and integration with the existing systems is a long-term process, and operational and strategic reviews must also be built in to the program.

19.10 APPLICATIONS OF EDI

EDI is most commonly associated with standard commercial transactions and messages: orders, invoices, and acknowledgements. There are, though, two other large categories: technical data, in which large quantities of structured data are exchanged between, for example, a design company and a manufacturing company, and specialized transactions, such as payment instructions between banks.

Different applications have different requirements for standards, connectivity, speed, and security. Some applications nowadays also require interactive EDI—the ability to send an EDI message and receive an immediate response.

19.10.1 Retail and Wholesale Distribution

Large retailers were among the first to make use of EDI in its present form. As retailing has moved to larger units and larger chains of shops, control systems have improved. The introduction of bar-codes and laser scanners in the 1970s made a big difference, as it allowed a more structured approach to the whole problem of inventory management.

Retailers can benefit from all the standard advantages of EDI. But a key motivation for retailers to make use of EDI lies in the inventory savings and dynamic stock management that it allows. Many retailers are now using supply chain management techniques with the help of EDI. Fresh foods are an area in which the fast turn-around of the EDI transaction is a particular benefit.

The concentrated nature of retailing and wholesaling in many countries means that a few dominant companies, often the large supermarkets or clothing chains, can exert significant pressure on their manufacturers and other trading partners to adopt EDI or other business methods.

19.10.2 Transportation

EDI is used in all kinds of transportation system: in shipping and ports, airlines and airports, railways, and road transport. In the latter case, it is often linked to retail and wholesale distribution systems. EDI can offer better control of routes and driver hours, and can improve the utilization of vehicles by selling "empty legs" or optimizing consolidated loads.

International transport companies will benefit from the streamlined customs procedures now being introduced throughout Europe. The direct interface between users, transport companies, and customs authorities means, however, that the role of the freight forwarder, who acts as a pivot in many transportation transactions, will have to change.

19.10.3 Manufacturing

The shape of manufacturing industry has changed greatly in the last thirty years. Heavy industry has generally declined in all the western industrial countries. The automobile industry has had mixed fortunes, but it has changed its habits in many ways, and is now committed to new methods of production and to new techniques such as EDI. The ODETTE standards in Europe and the AIAG standards in North America are among the most important single-sector EDI standards.

The output of the defense and aerospace industries has fallen very sharply in the last few years, but there is a strong commitment to systems in these industries, and organizations such as NATO and the U.S. Department of Defense have taken a strong line in promoting the use of standards compatible with EDI. The electronics industry is a major user of its own products, including EDI.

Manufacturers who adopt EDI also find it much easier to move to new production techniques such as just-in-time manufacturing, flexible manufacturing, and simultaneous engineering. In some cases, these can dramatically improve the performance of the plant as well as reducing costs.

19.10.4 Electronic Funds Transfer

EDI is simply a necessity for a modern bank clearing system. It is unthinkable that every transaction should be entered twice, once by the paying bank and once by the payee bank.

Interbank EDI is therefore well established, initially using private networks and proprietary protocols, but now starting to use EDIFACT standards in a limited way. Commercial banking EDI is also getting under way, with some small-scale systems that allow customers to send payment instructions to their bank online or to perform a limited range of other transactions or inquiries.

One of the largest EDI systems in existence is the credit card clearing system. This is also under review at present, and is likely to be upgraded and made more internationally compatible in the next few years.

19.10.5 Other Applications

For an industry sector to make use of EDI, it must have a reasonable penetration of computer systems, a generally accepted way of doing business, and a common terminology. A sector is most likely to introduce EDI where the competitive structure includes a moderate number of fairly large firms: too many small companies will not have the coordination, while one or two very large firms will not see it in their interests to promote an industry initiative.

Examples of other EDI applications are to be found in the insurance and securities industries, in travel and construction, in the oil industry, and in the public sector.

19.11 EDI SERVICES AND NETWORKS

The largest international EDI networks are offered by AT&T EasyLink, BT Tymnet, GEIS (which includes the UK-based INS) and IBM. All of these also offer other network services, such as electronic mail, to their subscribers. The EDI services include the provision of software.

There is some competition to these services from the public networks in some countries. At present, most EDI application users find it more attractive to run on the private VANs, which offer extra services in return for extra cost; however, the level of service available from the public networks is increasing.

19.12 THE INTERNATIONAL SCENE

Most EDI takes place within national borders. This is largely because most trade is domestic, but there are also organizational and technical problems in exchanging data between two countries, even where an international standard exists. EDI between the offices of a single company is often the exception to this.

With the growth in customs EDI systems in Europe, international trade is expected to develop into a major application of EDI in its own right.

North America, Europe, and Japan all have very different characteristics when it comes to EDI. In North America, the emphasis has been on large, horizontal-market, national schemes, often sponsored by a government agency. In Europe, where the majority of the initiatives have been initiated by commercial pressures (but sometimes developed under a European Community scheme, such as TEDIS), smaller and more closely defined user groups have been commoner. In Japan, EDI has been driven by closed user groups, often sponsored by a major manufacturer.

19.13 CONCLUSIONS

It is now time to draw some conclusions from this review.

The direct benefits of EDI—the savings in time and money that can be realized—are real but the competitive advantage that a company can derive from them may be short-lived.

The most obvious problems with EDI, such as legal issues, security, standards, and personnel-related problems, can easily be overcome, although they must not be ignored.

Issues, such as the responsibility for holding inventory, reflect traders' perceptions of the customer-supplier relationship: they are not so much one-sided benefits as an opportunity for service.

A mature view of EDI sees it as a way to optimize the business process and to offer service to the customer in the most cost-effective way possible. EDI must be taken into account in determining the strategy of the business.

19.14 A CHECKLIST

To finish up, here is a checklist of areas that a strategic review might take into account in making a business decision about EDI. Can we use EDI to:

- Reduce communications or postage costs?
- Reduce financial administration?
- Be paid more promptly?
- Self-bill based on EDI documents?
- Offer a better service to customers?
- What better service would they want?
- Reduce stock?
- Reduce time spent sorting out mistakes?
- Free order-input or sales-administration staff to deal with exceptions rather than routine tasks?
- Receive better management reports?
- Provide better information for ordering or purchasing?
- Institute better job control or planning?
- Let operations or production know *why* an order is urgent?
- Prioritize and introduce *degrees* of urgency?

- Implement a lean production program?
- Reduce stock waste or needless work-in-progress stock?
- Gain new customers?
- Introduce new services or business lines?
- Avoid our competitors gaining from any of these?
- Make ourselves less prone to the ups and downs of the business cycle?

Appendix

Addresses of Principal Network Suppliers in Europe

AT&T EasyLink Services
4 Moons Park
Burnt Meadow Road
North Moons Moat
Redditch
Worcs B98 9PA
England

British Telecom (GNS)
Network House
Brindley Way
Hemel Hempstead
Herts
England

GE Information Services Ltd
4th Floor
Shortlands
London W6 8BX
England

IBM Information Exchange
International Market Support Centre
PO Box 60
2700 AB Zoetermeer
The Netherlands

International Network Services Limited
INS House
Station Road
Sunbury-on-Thames
Middx TW16 6SB
England

Index

The Artech House Telecommunications Library

Vinton G. Cerf, Series Editor

Numerical Analysis of Linear Networks and Systems, Hermann Kremer *et al.*

Optimization of Digital Transmission Systems, K. Trondle and Gunter Soder

The PP and QUIPU Implementation of X.400 and X.500, Stephen Kille

Packet Switching Evolution from Narrowband to Broadband ISDN, M. Smouts

Principles of Secure Communication Systems, Second Edition, Don J. Torrieri

Principles of Signals and Systems: Deterministic Signals, B. Picinbono

Private Telecommunication Networks, Bruce Elbert

Radiodetermination Satellite Services and Standards, Martin Rothblatt

Residential Fiber Optic Networks: An Engineering and Economic Analysis, David Reed

Setting Global Telecommunication Standards: The Stakes, The Players, and The Process, Gerd Wallenstein

Signal Processing with Lapped Transforms, Henrique S. Malvar

The Telecommunications Deregulation Sourcebook, Stuart N. Brotman, editor

Television Technology: Fundamentals and Future Prospects, A. Michael Noll

Telecommunications Technology Handbook, Daniel Minoli

Telephone Company and Cable Television Competition, Stuart N. Brotman

Terrestrial Digital Microwave Communciations, Ferdo Ivanek, editor

Transmission Networking: SONET and the SDH, Mike Sexton and Andy Reid

Transmission Performance of Evolving Telecommunications Networks, John Gruber and Godfrey Williams

Troposcatter Radio Links, G. Roda

Virtual Networks: A Buyer's Guide, Daniel D. Briere

Voice Processing, Second Edition, Walt Tetschner

Voice Teletraffic System Engineering, James R. Boucher

Wireless Access and the Local Telephone Network, George Calhoun

For further information on these and other Artech House titles, contact:

Artech House
685 Canton Street
Norwood, MA 01602
(617) 769-9750
Fax:(617) 762-9230
Telex: 951-659

Artech House
6 Buckingham Gate
London SW1E6JP England
+44(0)71 630-0166
+44(0)71 630-0166
Telex-951-659